ENVIRONMENTAL PUBLICS

How do ordinary people think about the environment as they go about their daily lives? Does thinking about the environment make them do things differently? This book is the first to explore the idea of 'environmental publics', that is, the ways in which ordinary people engage with environmental issues across different practical contexts of work, play and home.

Emphasising the practices of 'environmental engagement', *Environmental Publics* examines how people consume the environment, learn about it, campaign for its protection and enjoy it through their leisure time. But the book avoids relying on idealisations of 'consumers' or 'citizens', or theoretical constructs about behavioural norms that have traditionally dominated research in this field. Instead, this book differentiates environmental publics not by who they are but by what they are doing – their daily practices. It also analyses specifically the geographies of those practices – how what people do affects the environment but in different ways across time and space and at different scales – aspects of practices that are neglected in the literature.

With an interdisciplinary perspective, this book will be of interest to students and scholars in geography, sociology, science and technology studies, political science and anthropology. It is written in an accessible and readable style, so as to be useful for preliminary and more advanced courses in environmental management, perception and policy, as well as in studies of modern society, consumption and environmentalism.

Sally Eden was a professor in the Department of Geography, Environment and Earth Sciences at the University of Hull, UK. She spent twenty years researching how people relate to the environment through consumption, leisure, knowledge and policy.

ENVIRONMENTAL PUBLICS

Sally Eden

Routledge
Taylor & Francis Group

LONDON AND NEW YORK

First published 2017
by Routledge
2 Park Square, Milton Park, Abingdon, Oxon OX14 4RN

and by Routledge
711 Third Avenue, New York, NY 10017

Routledge is an imprint of the Taylor & Francis Group, an informa business

British Library Cataloguing in Publication Data
A catalogue record for this book is available from the British Library

Library of Congress Cataloging in Publication Data
Names: Eden, Sally, author.
Title: Environmental publics/Sally Eden.
Description: Milton Park, Abingdon, Oxon; New York, NY: Routledge, 2017. |
Includes bibliographical references and index.
Identifiers: LCCN 2016032226 | ISBN 9781138189409 (hardback: alk. paper) |
ISBN 9781138189416 (pbk.: alk. paper) | ISBN 9781315641591 (ebook)
Subjects: LCSH: Human ecology. | Nature–Effect of human beings on. |
Environmental protection–Citizen participation.
Classification: LCC GF41 .E34 2017 | DDC 304.2–dc23
LC record available at https://lccn.loc.gov/2016032226

ISBN: 978-1-138-18940-9 (hbk)
ISBN: 978-1-138-18941-6 (pbk)
ISBN: 978-1-315-64159-1 (ebk)

Typeset in Bembo
by Deanta Global Publishing Services, Chennai, India

CONTENTS

FIGURES

TABLES

ACKNOWLEDGEMENTS

This book is an attempt to bring together some of the various ideas that I have toyed with, tried out and developed over the last few years of writing separate articles about how people understand, engage with and change environments.

It benefits from the ideas and support of many people over the years. Some of the key ideas were developed when I worked on the 'Angling in the Rural Environment' project funded by ESRC/NERC/BBSRC under the Rural Economy and Land Use (Relu) Programme (RES 227 25 0002) and on the 'Credibility Claims as Scientific Commodities' project funded by the Economic and Social Research Council's Science in Society Programme (Phase 2: RES-151-25-0035). I am very grateful particularly to Chris Bear and Gordon Walker with whom I worked on those projects and for their continued friendship and collaboration afterwards.

Many thanks also to my 'critical friends' who provided comments on my work, especially Lewis Holloway, Carol Morris, Gordon Walker, Paul Barratt and David Gibbs who read chapters of this book in draft, providing many useful comments to improve the final version, as well as positive feedback. The departmental writing group in Hull was also a great motivator to complete some of the chapters, so thanks to all who attended those meetings and good luck with your own writing.

I also drew on my experiences outside academia for Chapters 3 and 6, especially on my experiences of working with neighbours and other campaigners against proposals for a new waste facility. I hope they will feel that I have represented their efforts accurately and I thank everyone who was involved in that campaign for their input and discussions.

My family have been incredibly supportive of me during the time it took to finish this book after several years of planning it. During several bouts of chemotherapy and other treatments, the writing of this book became a merciful distraction from hospital visits, physical discomforts and ongoing stresses. My parents, my children and my husband have all been wonderful in the last two trying years, so much love and thanks to them. My niece, Helena Long, provided some great illustrations to liven up the text and make this book a real family project, so many thanks again.

Staff at the Sir Robert Ogden Macmillan Centre at Harrogate District Hospital have also given me unstinting physical, mental and emotional support over the last six years of living with cancer. That is why I have decided to donate any royalties from this book to the Centre, to help them, even in a very small way, to continue to help others to live their lives as fully as possible.

1

DIFFERENTIATING ENVIRONMENTAL PUBLICS

Introduction

Introduction

What do we mean by 'the public'? The term is used all the time, in newspapers and everyday talk, in academic analyses and in policy exhortations, yet it is often not explicitly defined or differentiated. Yet what we mean by 'the public', in phrases like 'public space' and 'public participation', directly affects many democratic processes and the framing and success of political struggles over policy and social change.

And being clear about what we mean by 'the public' is particularly important when we talk about involving 'the public' in environmental policy and practice. Success in implementing policy often depends upon public awareness and action, whether in terms of designing recycling systems that can be easily incorporated into people's everyday lives, or getting the public to support environmental protection and prevent or adapt to climate change. So it is important to be explicit about what we mean when we talk about 'the public' in connection with environmental policy and action and how this affects the framing and success (or otherwise) of that policy and action.

To do this, I want to argue in this book that 'the public' is not one group, but a highly diverse and complex set of groups that vary by context in time and space. There is not one mass or general public to which environmental policy can turn, but many – what I call 'environmental publics'. And these publics are not differentiated by their age, gender, income or education – what are called their 'sociodemographics' when the public is analysed and classified into 'segments' for policy use (e.g. Defra 2008; Natural England 2012, p. 16) – as has often been assumed. Instead, environmental publics are differentiated by how they relate to the environment through their environmental practices, rather than their own characteristics.

From an academic perspective, this raises several theoretical issues, such as how to differentiate publics and why differentiating publics matters. For example,

Staeheli et al. (2009) critiqued the notion of 'the public' and argued that it would be more accurate to see political struggles over public recognition producing fragmented, multiple publics that are unstable and that shift and change over time. And Irwin (2006, p. 314) considered the "multiple constructions of the public" at play in public engagement debates about GM food and the biosciences. As I shall explore later, these constructions have both good and bad interpretations – some publics are seen as 'good' (that is, useful, democratic and representative), whereas other publics are seen as 'bad' (that is, disruptive, undemocratic and driven by extreme opinions and narrow self-interest). These interpretations can then have important effects on how those publics are involved (or not) in a range of environmental debates and therefore the power or influence that they can exert.

However, most previous work thinks about publics in terms of how they are addressed or 'spoken to', particularly through public engagement exercises organised by government. In this book, I want to look at how multiple constructions of the public develop and are performed in other ways beyond speaking and writing about them and I will do this by attending more explicitly to a wider range of the different environmental *practices* involved in (and beyond) traditional participation exercises. I use the term 'environmental engagement' to encompass both ways of *knowing* and ways of *doing* in relation to the environment, precisely because this is a more inclusive term than other more common alternatives such as 'environmental perception' or 'public participation'. Importantly, 'engagement' encompasses both knowing and doing, being and behaving, imagining and acting, thus ranging across diverse practices, both discursive and embodied.

Publics in environmental practice

By focussing upon what publics do and how this affects how publics are understood, studied and engaged with by policy makers and other elite groups, I am drawing on what I have learnt from my own research projects over the last fifteen years, studying how people relate to the environment in quite different ways, for example, as consumers and as recreationists.

But this book also reflects what has been called 'the turn to practice' in social science (Schatzki 1996; Whatmore 2006) in recent years, particularly in relation to consumption, science and technology. This emphasis on *practice* has been used in different ways. For example, in the sociology of consumption, Warde (2005) argued that theories of practice had the potential to revive analyses of consumption in everyday life, drawing on Schatzki (1996) and Reckwitz (2002). In consumption studies relating specifically to the environment, Shove (2010, p. 1279) argued that the 'catch-all' list of differentiating 'factors' that have been used to try to explain environmental behaviours and predict how they will change have failed to capture many important aspects of social change, because this approach has neglected practice: "understanding social change is in essence a matter of understanding how practices evolve, how they capture and lose us, their carriers, and how systems and complexes of practice form and fragment".

Recently, Warde (2014) argued that practice theories (which he saw as diverse, hence the plural) should be used to counter the model of consumers as individual but rational agents that had come to prevail since 'the cultural turn' in sociologies of consumption. Similarly, Spaargaren (2011) argued for a practice-based approach to overcome the deficiencies of analyses of environmental change based *either* on individual agency *or* on structural power. Both authors were arguing for a more collective, more materially based approach to analysing consumption and consumers. "Practices, instead of individuals, become the units of analysis that matter most" in Spaargaren's (2011, p. 815) view, and it is practices that produce individuals, not individuals that produce practices. Practice theories, therefore, emphasise

> routine over actions, flow and sequence over discrete acts, dispositions over decisions, and practical consciousness [see Giddens 1986] over deliberation ... doing over thinking, the material over the symbolic, and embodied practical competence over expressive virtuosity in the fashioned presentation of self.
>
> *Warde 2014, p. 286*

As well as being diverse, practices are themselves complex assemblages of diverse elements and dimensions that are put together and used in different ways.

> Practice consists of three basic elements: stuff (both nature and objects); image (symbols and meanings); and skill (competence, know-how and technique). Practice-as-entity is held together by these heterogeneous elements, which are linked by practitioners when practices are performed. In this vein, practices exist, persist or disappear when the links between these three elements are created, sustained or broken.
>
> *Truninger 2011, p. 40*

Hence, and importantly, we need to link practices with not only people, but also with objects. And I refer to 'practices' rather than 'social practices' in this book precisely because many environmental practices that I shall consider are not merely social; instead, they are 'more than human' in how they are performed, relying on technology and (other) animals, as well as water, land and air.

Practice theory has clear parallels with actor-network-theory (ANT) approaches from the sociology of science and technology, as both approaches emphasise how active human and nonhuman agents interact and shape each other to co-produce worlds. Although ANT approaches tend not to use the term 'practice' in a generic sense, Law and Mol (2002) do use it when looking at how knowledge shapes practices and practices shape knowledge. For instance, designing an airplane embeds engineering knowledge into routinized practices, such as wing design, so that eventually that knowledge is 'black-boxed', that is, hidden away and unquestioned, although the practices continue to perform the design task that it enabled (Law 2002).

What is useful about practices for analysing environmental publics specifically? First, practices are shared but not homogeneous, so we can analyse how people do things together (birdwatching, recycling, campaigning against an incinerator) but remain diverse in their sociodemographics, rationales and roles. For example, when I was studying recreational anglers, if we brought together five anglers in the same room to discuss an environmental issue such as water pollution or climate change, we usually heard five different opinions. In this sense, a focus on practices is relational (Halkier and Jensen 2011), but not universalising: recognising diversity amongst environmental publics and differentiating practices means that we cannot conceive of a single environment, a single nature 'out there', but multiple "natures" (e.g. Macnaghten and Urry 1998) produced by and through different practices and beliefs.

Second, practices are repeated and thus are not merely acts or 'behaviours' (Shove 2010) that are one-offs or autonomously generated by some underlying individual attitudes or choices (Halkier and Jensen 2011); rather, practices are produced in and through contexts of shared relationships, infrastructural constraints and habits.

Third, through repetition, practices are performative of everyday social life, of time and space and of present and future worlds. Practices are continually enacted (Schatzki 1996) or 'routinised' and thus regularly reproduced (Reckwitz 2002), becoming part of the social fabric of our lives, shaping our expectations and social norms. In this sense, practices are carried by people through regular (but always slightly different) performances over time, becoming more or less stable over time, but also being capable of change. Shove and Pantzar (2005, p. 62) argued that practices co-evolve with practitioners who link together materials, skills and meanings; practices thus travel and grow in complexity through reinvention from place to place, from person to person, but do not merely 'diffuse' alone without people and without change. Waterton (2003, p. 114) argued that the performers of practice also refashion it and through this refashioning also refashion their social worlds, as "performances are improvisatory, situated, and, importantly, embodied, encompassing much more than cognition and intellect, thus attributing creativity to what may appear simply to be acts of replication and tradition". And through this performance, different (social and environmental) worlds come into being and environmental publics themselves are re-made, re-imagined and re-constituted, as the ways in which they engage with environmental issues change.

But considering the routinisation of practices can also cause us to lose sight of the way in which the world is made, as these performances are normalised and the choices within them (of transport modes, of types of foods, of disposal outlets) become naturalised, that is, hidden away behind tacit knowledge, codifications and other standardisations that make new practices into old habits. This means that practices sometimes become increasingly difficult to challenge and, where practices are not clearly articulated, very difficult to unthink or to think about doing differently. People may not realise that they are doing them at all or, if they do, they may not realise why they do them in particular ways and in some cases (such as when they are asked by researchers to explain why they buy a particular brand of breakfast food, for example) they rationalise their practice verbally after the fact quite cogently but possibly quite incorrectly, because words can express many practices only partially. This is particularly important when we study environmental publics, because often they have routinized unsustainable or environmentally damaging practices, which then become very difficult to re-think and change.

Theorising this process is, however, problematic. Both actor-network theory (ANT) and social practice theory expose this process of making (variously referred to as co-production, performance, translation), but often only by identifying more performances that precede it – there is no starting point, no ultimate cause, that they can identify. There are similarities here with ANT's understanding of how processes, devices and inscriptions stabilise, persist or (to use Latour's [1991] phrase) are 'made durable'. Latour (1986, 1991) argued that instead of thinking of power as the cause of action or as an essential attribute of a person or thing, power should be treated as a consequence, effect or product of collective action (see also Law 1991) and causal structures therefore do not pre-exist activity. By reorienting our picture of how the world is made, what Latour (1986 p. 273, emphasis in original) identified is that the "shift *from principle to practice* allows us to treat the vague notion of power not as a cause of people's behaviour but as the consequence of an intense activity of enrolling, convincing and enlisting".

Conceptually speaking, we might draw an analogy between practices and genes. A person as a performer is a carrier of practices that move and are shared with other people. A person – as a human body – is also a carrier of genes that move between generations by sharing and networking. Genes have different effects in different bodies and are also reproduced and/or changed (evolved) by the conditions of their enactment. Thus there is a continual re-making of people and practices through the interaction between genes and the specific context in which those genes are expressed or activated. But unlike genes, practices are less clearly isolated and defined under the microscope of social science – often practices blur into one another and shift and change surprisingly quickly.

Practices and power in place

Finally, I want to emphasise that practices are performed differently in specific places and spaces: the way that someone eats, uses a computer or learns about environmental change may differ depending on whether they are in their work office, their home or halfway up a mountain on a walking holiday. Practices can also travel, especially through writings and rules of different kinds, as well as through learning and networking, both person-to-person and through other forms of communication, such as using the internet.

This means that we need to attend not only to bundles of practices and their routinisation, but also to the *geographies of practice*. The geographical contextualisation and remaking of practices have been somewhat neglected in practice theory and ANT. In this book, I will particularly focus upon what each bundle of practices and their context contributes to the *spatialization* of environmental publics, including where practices take place, how they are scaled, their imagined geographies of influence and how they perform geographical and topological connections. In this, "scale is what actors achieve" through collective action (Latour 2005, p. 184), that is, it is constructed through acting, rather than pre-existing. Moreover, as well as making publics through environmental engagement practices, places are also made, both negatively through damage (e.g. overuse) and also positively through guardianship (e.g. rebuilding paths, constructing river banks, planting trees, picking up litter, lobbying for environmental protection and re-imagining spaces and their use in more discursive ways).

As a consequence of all of these aspects, practices matter materially and politically. It is perhaps obvious to say that how we think about and behave in relation to the environment – how we engage with it through more or less environmentally damaging ways – is often quite different, for example, when we recycle products at home compared to when we buy birthday gifts or manage teams at work. But just because something is obvious does not mean it is any less significant and policy often does not address these very obvious differences, tending instead to homogenise the public, as I shall show later in this book, frequently in quite negative ways that gloss over differences and lead to problematic assumptions that ignore context.

To emphasise this approach to thinking about the *geographies of practices*, I have divided practices up by context as well as by type in the chapters that follow. Different chapters consider practices that involve consuming in the home and consuming in the office, but within each chapter I also explore how practices are spatialized through literally taking place in different contexts. For example, people may learn about environmental change in the very environment that is changing (e.g. on a school field trip) but may also learn about it elsewhere (e.g. while in a classroom or watching TV in their living room).

In some cases, practices are organised by people (both publics and elites) by scale. Scalar organisation spatializes practices through ordering, prioritising and linking different scales of practices in a more generic sense, invoking scalar hierarchies when imagining, making and legitimating environmental publics. For example, Chapter 6 shows how groups campaigning for environmental reform may perform national networking quite differently from how they perform local networking and Chapter 7 shows how voters may make very different ballot decisions in local elections compared to national elections. So the scaling of practices differs from placing of practices, with the latter being more specific to context.

As well as looking at the spatialization of specific practices, I will also consider their consequences for *power*, that is, how powerful each is in terms of influencing and environmental impact, but also how that power (or lack of power) is imagined to be held by, or ascribed to, 'the public' in each context. As we shall see, this throws up some strong contradictions in the influence and legitimacy of different environmental publics, especially when it comes to claiming local environmental knowledge or protests by local people about environmental change in their neighbourhoods.

I have also explicitly chosen to use the word 'power' rather than the word 'politics' here. This was partly to avoid the ways in which ANT has been criticised

(Latour 2005, p. 251, says "accused") of extending politics to everything and everywhere (e.g. non-human animals and technologies), but also to avoid the word's association with traditional political spaces of public engagement, such as electoral hustings or public meetings organised by local government. Traditional political spaces will feature a little in some chapters of this book, especially Chapters 6 and 7, but many chapters will instead look outside these traditionally 'political' spaces to emphasise the more mundane, overlooked and contradictory politics that is performed by environmental publics in other spaces, performances that stitch places together through and across diverse practices.

Disciplinary differences

As well as dealing with practices and spatialisation, this book has an interdisciplinary flavour. This is because academics from different disciplines have tended to look at environmental publics and their practices using different theories and perspectives – and also sometimes different methods and measurements (e.g. opinion polls versus focus groups, see Macnaghten and Urry 1998). This means that our understanding of environmental publics is fragmented and sometimes contradictory, scattered across (sub)disciplines and interpreted in ways that over-emphasise some elements and patterns at the expense of others and exclude others completely.

These interpretations matter because they *do work* in the world. By that I mean that they are used to shape material practices, forms of consultation, language used and opportunities for public interventions. Our assumptions about 'environmental publics' thus have effects reflexively on the environment and society, including how we position ourselves as part of those publics. For example, how we understand (and often measure) environmental publics will shape how we design campaigns to change people's behaviour in relation to the environment and who (and where and when) we include in debates about the environment and people's relationship with it. In this sense, practices of defining and in/excluding various publics can enable or curtail the practices made available or attractive to those publics.

So how we define the public affects (a) democracy through public participation (or lack of it) in decision-making, (b) feasibility in terms of public action in policy implementation (e.g. consuming) and (c) legitimacy in terms of public support, especially for environmental protection by the state (e.g. voting). To put this another way, how we define publics also *performs* those publics and runs the risk of implicitly reinforcing the exclusions and presumptions inherent in the term 'environmental publics' without scrutiny and thus without redress.

These arguments have been raised in more general terms in connection with the 'public understanding of science' and deliberate efforts to increase public awareness and knowledge of technological developments to involve publics more in decisions about those developments. In such public engagement exercises, "publics are not merely being encouraged to express their citizenly concerns, they are also being 'made' as particular types of citizens by virtue of the models of the public that inform public engagement" (Michael 2009, p. 619), demonstrating how organising can make 'the public' in specific ways (Felt and Fochler 2010).

The ways in which the state, commercial companies and NGOs approach 'the public', both practically and through imagined and idealised models, shape those publics and give them particular roles to play (or reject). Those publics also perform their roles based on idealised models of themselves – what practice theorists like Shove referred to in terms of images, symbols and meanings – sometimes shared with other groups, sometimes quite differentiated. Through all these mechanisms of public engagement, whether through state-sponsored participatory decision-making, advertisements for sustainable products or worker environmental awareness schemes, publics are being made, performed and changed in various ways.

Sometimes, the very social science methods used to measure variables can define 'the public': survey questionnaires may limit them in terms of sociodemographic variables or by asking about what they know or do not know (e.g. Irwin and Michael 2003; Michael 1998; Wynne 1995). Methods thus build powerful images or constructs that then circulate in debates and frame how consultation and decision-making are organised in practice. In this sense, social science itself and its methods not only observe but can also create realities (Law 2008). But so far, social science literature has dealt little with how these ideas apply specifically to *environmental* publics, which has prompted me to write this book to bring these ideas together.

The structure of this book

Following the argument suggested above, this book has been organised so that environmental publics are differentiated through their practices of environmental engagement, rather than through their sociodemographics. The chapters that follow will each focus upon one set or context of practices that involve and shape environmental publics, such as participating, voting, working and consuming. Sometimes, specific disciplinary perspectives focus on analysing particular practices or contexts (e.g. voting); but for many practices, adopting only one disciplinary perspective hinders analysis. So many chapters of this book will draw on several different disciplinary (and subdisciplinary) perspectives, such as human geography, sociology, political science, psychology and planning, contrasting their implications for how we understand the practices, geographies and power of environmental publics. Each chapter in this book will analyse the four 'P's:

- Practices – what environmental publics do in a particular context and how they do it;
- Proportion – what publics are involved or implicated in those practices and how they are measured (often inconclusively);

- Places – the geographies of practices, the spatialisation of environmental publics and the contradictions involved;
- Power – how much those publics influence environmental change, either directly through management and policy or more indirectly.

Table 1.1 summarises how these four aspects of environmental publics operate in each chapter. This (very brief) summary outlines the contents of this book but also necessarily simplifies a great diversity of environmental practices and their associated environmental publics. Most chapters draw on my own research of diverse publics over the years, to illustrate the very diversity, the multiplicity encapsulated in the title of this book.

Chapter 2 looks at perhaps the most well-defined area of research – practices of environmental knowing amongst publics, such as formal education, self-education, knowledge production through perception and embodiment. The geographies of environmental knowing link together *in situ* and *ex situ* learning, as well as scalar hierarchies that are re-invoked to venerate knowledge as general or global, or vilify it as (merely) particular or local. Chapter 3 considers practices of participating in formal, usually state-sponsored systems of environmental decision-making and policy formation and the democratic norms associated with these practices. Again, scalar organisation is linked with vastly different evaluations of environmental practices, from denigrating local networking as NIMBYism (Not In My Backyard) to valuing local people's input to democracy. Chapter 4 concentrates upon practices of green consuming, especially through individual purchase, use and disposal, and how these are spatialized. It also examines how critics have denigrated consumption practices as less worthy, less democratic and less powerful than other, more citizenly practices.

Chapter 5 considers practices of environmental leisure, recreation and enjoyment, such as walking, fishing, swimming and watching nature. Sometimes also criticised as a form of consumption, such practices are also often codified by recreational clubs and embodied through physical engagement in the environment, emphasising personal geographies of encounter. Chapter 6 looks at practices of green campaigning and political activism, through both large and small non-governmental organisations, and considers the democratic consequences of such forms of public organising and how these are spatialized, such as through the scalar ranking of international and national networking over local efforts. Chapter 7 looks at practices of green voting, contrasts these across electoral settings to demonstrate the marked geographies involved, and considers how these compare to other sorts of political practices in terms of the moral worth ascribed to them as citizenly activities.

Chapter 8 considers practices of environmental working and how these are influenced by the control that is perceived to reside with the employer, and also the geographies by which they are developed, shared and travel within and between organisations. Finally, Chapter 9 re-emphasises the importance of geographies of practice, considers how practices are linked between different contexts in which

TABLE 1.1 Approaches to environmental publics studied in this book

	Practices	Proportion	Place	Power
Chapter 2: Knowing publics	Formal education, informal and self-education, media use and amateur knowledge practices	Minority according to some, but potentially majority or all	*In situ* and *ex situ* learning. Local knowledge valued, but general/universal knowledge presumed weak, lacking or incorrect	Variable, but often low, especially on issues seen as highly technical
Chapter 3: Participating publics	Attending meetings, being surveyed, learning, decision-making	Minority	Often local, but also national experiments	Potentially high if decisions remain open, but low if consultation is for legitimation rather than influence
Chapter 4: Consuming publics	Purchasing, using and disposing of things and services	Nearly all (depending on income)	Often not spatialized and explicitly distanced from sites of production, but also 'buy local' campaigns	Highly variable interpretations, from very low (according to political economy) to very high (according to business studies)
Chapter 5: Enjoying publics	Leisure activities outdoors, such as leisure, including walking, riding, surfing, swimming, and associated indoor practices, such as watching films	Minority in specialised activities, e.g. surfing, but majority in common practices, e.g. walking	Frequently local, but also distant in the form of tourism	Often disregarded, except where clubs organise campaigns for access to environments, but also hands-on lay management of local environments
Chapter 6: Campaigning publics	Organising, mobilising and protesting, online and offline and learning how to do so	Minority	All scales, but often spatialized as local	Variable, but often influential, especially in opposition, although accusations of NIMBYism may undermine influence
Chapter 7: Voting publics	Listening to, standing in and voting in formal elections	Small minorities vote 'Green' and many registered voters do not vote at all	Greatly influenced by scale, with local, national and supranational (EU) elections showing very different results	Low, due to low numbers of 'Green' votes
Chapter 8: Working publics	Knowing, consuming, producing, retailing, servicing, travelling etc. but in work context	Majority (depending on employment rates)	Potentially all scales, but often local or organisational as the focus	Potentially important but often neglected in policy

publics work, live and play, including online and offline spaces and addresses the vexed question of how environmental publics and their practices change over time, as well as space.

References

Defra, Department for the Environment, Food and Rural Affairs (2008). A framework for pro-environmental behaviours. Defra, London. www.gov.uk/government/publications/a-framework-for-pro-environmental-behaviours

Felt, Ulrike and Maximilian Fochler (2010). Machineries for making publics: inscribing and de-scribing publics in public engagement. *Minerva* 48, 219–238.

Giddens, Anthony (1984). *The constitution of society.* Polity Press, Cambridge.

Halkier, Bente and Iben Jensen (2011). Methodological challenges in using practice theory in consumption research: examples from a study on handling nutritional contestations of food consumption. *Journal of Consumer Culture* 11, 1, 101–123.

Irwin, Alan (2006). The politics of talk: coming to terms with the "new" scientific governance. *Social Studies of Science* 36, 2, 299–320.

Irwin, Alan and Mike Michael (2003). *Science, Social Theory and Public Knowledge.* Open University Press, Maidenhead.

Latour, Bruno (1986). The powers of association. 264–280 in John Law (edited), *Power, Action, Belief: a new sociology of knowledge.* Routledge & Kegan Paul, London.

Latour, Bruno (1991). Technology is society made durable. 103–131 in John Law (edited), *A Sociology of Monsters: essays on power, technology and domination.* Routledge, London.

Latour, Bruno (2005). *Reassembling the Social: an introduction to actor-network-theory.* Oxford University Press, Oxford.

Law, John (1991). Power, discretion and strategy. 166–191 in John Law (edited), *A Sociology of Monsters: essays on power, technology and domination.* Routledge, London.

Law, John (2002). On hidden heterogeneities: complexity, formalism, and aircraft design. 116–141 in John Law and Annemarie Mol (edited), *Complexities: social studies of knowledge practices.* Duke University Press, Durham.

Law, John (2008). On sociology and STS. *The Sociological Review* 56, 4, 623–649.

Law, John and Annemarie Mol (2002). Complexities: an introduction. 1–22 in John Law and Annemarie Mol (edited), *Complexities: social studies of knowledge practices.* Duke University Press, Durham.

Macnaghten, Phil and John Urry (1998). *Contested Natures.* SAGE, London.

Michael, Mike (1998). Between citizen and consumer: multiplying the meanings of the "public understanding of science". *Public Understanding of Science* 7, 313–327.

Michael, Mike (2009). Publics performing publics: of PiGs, PiPs and politics. *Public Understanding of Science* 18, 5, 617–631.

Natural England (2012). Monitor of engagement with the natural environment: the national survey on people and the natural environment. Annual report from the 2011–12 survey. Natural England, London. www.naturalengland.org.uk/ourwork/enjoying/research/monitor

Reckwitz, Andreas (2002). Towards a theory of social practices: a development in culturalist theorizing. *European Journal of Social Theory* 5, 2, 243–263.

Schatzki, Theodore R. (1996). *Social Practices: a Wittgensteinian approach to human activity and the social.* Cambridge University Press, Cambridge.

Shove, Elizabeth (2010). Beyond the ABC: climate change policy and theories of social change. *Environment & Planning A* 42, 1273–1285.

Shove, Elizabeth and Mika Pantzar (2005). Consumers, producers and practices: understanding the invention and reinvention of Nordic walking. *Journal of Consumer Culture* 5, 1, 43–64.

Spaargaren, Gert (2011). Theories of practices: agency, technology, and culture. Exploring the relevance of practice theories for the governance of sustainable consumption practices in the new world-order. *Global Environmental Change* 21, 813–822.

Staeheli, L.A., Mitchell, D. and Nagel C.R. (2009). Making publics: immigrants, regimes of publicity and entry to "the public". *Environment and Planning D: Society and Space* 27, 633–648.

Truninger, Monica (2011). Cooking with Bimby in a moment of recruitment: exploring conventions and practice perspectives. *Journal of Consumer Culture* 11, 37–59.

Warde, Alan (2005). Consumption and theories of practice. *Journal of Consumer Culture* 5, 2, 131–153.

Warde, Alan (2014). After taste: culture, consumption and theories of practice. *Journal of Consumer Culture* 14, 279–303.

Waterton, Claire (2003). Performing the classification of nature. 111–129 in Bronislaw Szerszynski, Wallace Heim and Claire Waterton (edited), *Nature Performed: environment, culture and performance*. Blackwell, Oxford.

Whatmore, S. (2006). Materialist returns: practising cultural geography in and for a more-than-human world. *Cultural Geographies* 13, 600–609.

Wynne, Brian (1995). Public understanding of science. 361–388 in Sheila Jasanoff, Gerald E. Markle, James C. Petersen and Trevor Pinch (edited), *Handbook of Science and Technology Studies*. SAGE, London.

2

KNOWING PUBLICS

Introduction

In a 1999 opinion poll, 35 per cent of respondents surveyed across Europe agreed with the statement "Ordinary tomatoes do not contain genes, while genetically modified tomatoes do", with 35 per cent disagreeing and 30 per cent saying that they did not know (INRA [EUROPE] – ECOSA 2000). In the UK specifically, even more people – 40 per cent – said that that they did not know. As part of the developing debate about genetically modified (GM) food at the time, many commentators used this piece of data (which remained fairly consistent across several years of surveying) to bemoan the state of scientific education in modern society and to argue for more public information about GM food, as they thought this would overturn public distrust of this new technology. Others pointed out that more education does not necessarily translate into more support for genetic modification or any other form of technology – research is inconclusive or contradictory about the effect of information provision on purchases of GM food, for example.

This chapter is about practices of environmental knowing amongst publics, which includes learning about science such as GM, but also other forms of knowing and learning about the environment. It is not about *what* people know, that is, the facts that they do or do not hold in their heads, but about *how* they come to know (or not know) about environments: the knowing and learning practices that environmental publics perform and what these teach us about geographies of environmental publics and the implications for power relations.

This chapter comes early in the book partly because knowledge about a problem tends to come before action to address that problem. But this does *not* mean that knowing is separate from or always comes before doing. In later chapters, I will show that often doing things is a process of learning as well, so that knowing can also come *during* or *after* doing as well. Conversely, sometimes people know something but do not apply their knowledge to their actions – the persistence of tobacco smoking despite widespread knowledge that it causes disease and death is a good example. Psychologists call the ability to think one thing but do its opposite without apparent conflict 'cognitive dissonance' and we will consider how this applies to environmental practices in later chapters when we look at how learning is also tied to participating, consuming and campaigning.

So, the link between knowing and doing is complex, often broken and more importantly moves in both directions. But a book has to start somewhere and be organised in some linear fashion, so I have put this chapter about knowing here, to concentrate on practices that are specifically and directly related to learning in some way. Later chapters look at practices that are more indirectly related to learning, such as learning about flooding while out having fun fishing or swimming, or learning about air pollution while campaigning against a local development project, practices that I will also mention in this chapter.

Having dealt with that caveat, we can now move to analyse how environmental publics become and are seen to be 'knowing' (and 'unknowing'), beginning with formal education and moving on to consider more informal channels, such as broadcast media, as well as public experiments in engagement. Implicitly also, levels (and kinds) of environmental knowledge amongst different publics will also shape the practices discussed in other chapters (e.g. purchasing, voting); again I want to emphasise that the divide between different practices is somewhat artificial, but it will serve to organise material in hopefully more readable ways and facilitate comparisons.

Knowledge-practices

Our first problem is how we define environmental publics as 'knowing' or not. This can be more difficult than defining consuming, recycling and voting publics in later chapters, all of which can at least be estimated in terms of money (pounds sterling), weight (tonnes) or votes cast. Despite this, many surveys have tried to measure how many people 'know' key environmental facts, such as the one in my opening vignette, usually using questionnaire surveys such as the regular Eurobarometer polls conducted across the member states of the European Union (see http://ec.europa.eu/public_opinion/index_en.htm). For example, in the 2011 Eurobarometer report specifically about *Attitudes of European Citizens towards the Environment* (p. 27), conducted by TNS Opinion & Social for the EU Directorate-General for the Environment, 60 per cent of respondents said they felt "fairly well-informed" (52 per cent) or "very well-informed about environmental issues" (8 per cent) with the rest feeling "fairly badly informed" or "very badly informed". These figures varied geographically, with the highest results reported from Sweden and Denmark, averaging 80 per cent "fairly well-informed" or "very well-informed", and the lowest in Bulgaria, Czech Republic, Romania, Spain and Portugal, reporting below 50 per cent.

But such measurements leave us with huge questions. What counts as 'informed' and which 'environmental issues' are people thinking about when they respond to such surveys? How accurate is the information that they (think they) have and how much do they trust it? Do they use such information at all, do they put it into practice in everyday life? In other words, we need to understand how people encounter, make sense of and judge information about environmental issues, that is, what knowledge-practices they perform and through which they become more (or less) knowing publics.

'Scientific literacy' is often taken as a marker of public understanding of science and how the world works generally, as in my opening vignette about genes in GM tomatoes. This means that, where environmental publics that have 'scientific literacy' are felt to have sufficient information about how science works and what the current scientific consensus is on key issues, so that they can make sense of scientific information in their everyday lives. 'Environmental literacy' is a variant of this scientific literacy and is often promoted through educational initiatives to foster understanding of how environmental systems work and, often but not always, how sustainable development and environmental protection can be implemented.

So knowing also performs environmental publics, for example, by shaping how people react to environmental stories in the media, what things they buy and use around the home, how they make sense of environmental changes over time and especially, as we are considering how environmental practices are performative of everyday life, through knowing also as part of doing and being, that is, how self-identifying as 'knowing' influences the multiple ways in which publics engage with the environment.

To unpack these knowledge-practices and how they produce environmentally 'well-informed' or 'ill-informed' publics, we can identify and attempt to measure 'knowing about environmental issues' first through formal education in school, college, university or other scientific or professional training, and second through more informal mechanisms.

Formal education

In many countries, knowing about the environment is laid down in curricula that explicitly define 'environmental education' or implicitly include it under other subjects taught at schools, colleges and other institutions of learning. But the content and rubrics of these curricula vary between different countries and between regions in countries, even down to the school level.

For example, the UK's National Curriculum in 2014 specified 'Geography' (including environmental change and sustainable development) as a statutory subject to be taught to all children in Key Stages 1–3 (roughly ages 5–14) and 'Science' (including living things in the local environment and effects of human activity on the environment) to be taught to all children in Key Stages 1–4 (ages 5–16). 'Citizenship' was also required to be taught at Key Stages 3–4 and is cross-referenced to geography through the concept of sustainable development and influencing decisions about the environment. The subject of 'personal, social and health education' also specifically deals with the process of public participation in decision-making about the environment, but was not a statutory subject for all children. In France, the National Curriculum (INCA 2012) requires all students to take 'History & Geography' and 'Civic, legal and social education' at ages 16–18, but the requirements to take science vary by the type of Bac chosen.

From this, we can see that codes developed by the state to support formal (class-based) education seem to make environmental publics in two different ways: (a) through teaching scientific or physical principles through environmental science and (b) through reflecting on and encouraging public, political and personal roles and actions in relation to environmental issues. These are taught in different settings, sometimes by different teachers (although Citizenship has often been taught by geography teachers in both the UK and France) and with different topics, although cross-referenced. This division between knowing and doing in practical terms about the environment is therefore offered to children very early in their lives, despite being very artificial and unrepresentative of more informal ways of learning, as we shall consider later. As O'Riordan (1981, p. 15) noted decades ago, "at the heart of environmentalism lies the practice of citizenship", which is assumed to support not only environmental (scientific) knowing, but environmental (political) doing as well (also Holsman 2001).

In terms of formal education, the UK shows little appetite for continuing 'citizenship' education for pupils after age 16 and environmental subjects remain a minority interest. If we take A levels as an example of qualifications for sixteen- to eighteen-year-olds in England, geography and biology are the most obviously environment-related subjects, but made up only 4 per cent and 7 per cent respectively of the total A levels taken in summer 2015 (801,510 entries), although geography entries had risen 14 per cent since the previous year.[1] The 2008 Strategic Review of sustainable development in UK Higher Education[2] broadened the definition of environmental education to include both natural and social aspects, and found the following subject categories to be the most commonly used "key words" related to sustainable development:

- Geography, Earth, Marine and Environmental Science;
- Architecture, Built Environment and Planning; Business and Management Studies;
- Sports and Leisure Studies;
- Civil Engineering.

But it also noted that there seemed to be little 'demand' from students or employers for sustainability-related courses to provide the learning and skills needed for what they called 'the sustainable development agenda' or what Wals and Blewitt (2010) referred to as 'sustainability competence'. Maybe these skills and knowledge are embedded in courses with other titles, from engineering to support low-carbon products to political science to support environmental policy, which the Strategic Review failed to pick up. But it seems that neither higher education institutions nor their funder, HEFCE, are pushing the creation of environmental publics through formal practices of higher education in the UK.

Although the HEFCE Strategic Review refrained from quantifying its conclusions, referring only to the great heterogeneity in how English higher education institutions define and deliver sustainable development through teaching, we can look at the numbers of students opting for degrees likely to contain sustainability and environmental components as quantitative (if very rough) proxies of public interest. Today, formal education reaches more people than ever before in history, with 26 per cent of the UK population in the 2011 Census having at least a Bachelor's degree or equivalent (Office for National Statistics 2014) compared to 23 per cent reporting no formal qualifications. Yet environmental education remains a minority interest, with only about 16 per cent (over 80,000) of students entering higher education in the UK in 2010 choosing courses that we could reliably expect to have some environmental components (Table 2.1.).

These measurements, although imperfectly defined, suggest that formal education in environmental subjects remains a minority interest, albeit a well-established one. But does such environmental education shape students into more environmental publics, both during their studies and afterwards when they enter jobs, start up households and so on? Evaluating the impact of environmental education on what people do the decades after formally completing their education

TABLE 2.1 Students choosing to begin environment-related undergraduate courses in UK higher education institutions in 2014

Higher education subject group (as classified by UCAS)	Students accepting a place on the course	Students accepting a place on the course as %
Biological Sciences	48,815	9.5%
Veterinary Science, Agriculture & related (including Forestry)	6,740	1.3%
Physical Sciences (including Chemistry and Geology)	19,920	3.9%
Architecture, Build & Plan (including Landscape Design)	7,685	1.5%
Total for these four subject groups only	83,160	16.2%
Total for all subject groups	512,370	100.0%

Source: Extracted from data provided by UCAS (2014) www.ucas.com/sites/default/files/eoc_data_resource_2014–dr3_010_01.pdf

Notes: UCAS has no subject names beginning with the word 'Environment' and only one containing the word 'environment' ('Science of Aquatic and Terrestrial Environments') and that is part of the Biological Sciences Group.

is tricky, but in a study of UK graduates, Whitmarsh (2009, p. 417) found that graduates (and especially science postgraduates) were more likely to be both "aware of scientific models of climate change and more skeptical that [climate change is] a human-caused problem" and therefore be less concerned about it. And in the USA, Zimmerman and Cuddington (2007) found that ecology students retained pre-existing understandings of 'the balance of nature' as ecological equilibrium, even though these were challenged explicitly by their formal training. Such studies suggest that more knowledge (or at least being closer to science through formal education) does not necessarily produce support for scientific consensus or concern about environmental damage amongst students.

And as well as educating students *about* the environment, much environmental education and 'education for sustainable development' also seeks explicitly to educate students *for* the environment, shaping their understandings and future practices at work, home and leisure. For example, the Eco-Schools initiative in the UK aims not only to raise awareness of environmental issues and sustainability amongst pupils, but to change their practices in schools and at home so as to reduce environmental damage overall. Here, learning is specifically and explicitly linked to shaping public conduct, under the assumption that information and knowledge can influence behaviour, an assumption that I shall challenge later.

In situ and ex situ ways of knowing the environment

First, I want to consider the inherent geographies in learning about the environment. For example, learning as environmental knowledge-practices can take place both *in situ* and *ex situ*, i.e. in the environment that is being studied (forests, seashore) or outside that environment (classrooms, homes). So far, I have looked mainly at *ex situ* learning in schools and universities, but directly experiencing the environment that is being studied is often seen as important for students. This is why fieldwork is seen as an important knowledge-practice in environmental education, particularly geography and biology, and many environmental sites (such as nature reserves and national parks) have developed detailed and extensive environmental education resources to be used *in situ* by groups visiting from educational institutions, as well as by families and other more informally organised groups visiting for pleasure.

For example, Figure 2.1. shows an information board about a nature reserve created out of a disused quarry in northern England, located so that the wetlands it describes are visible to the left of the visitor as they read the board. In other places, people rather than boards give environmental information and prompt learning. For example, park rangers in US National Parks often lead guided tours, explaining to visitors how the landscape was formed and how it is vulnerable to damage by people. This offers a learning experience that happens while the person experiences the very environment under study, an experience that is not only interactive because the visitor can ask the ranger questions as they walk, but also because the visitor experiences the physical sensations of being in that environment, such as disorientation, dusty heat or muddy pools after a downpour.

FIGURE 2.1 Information board at Nosterfield Nature Reserve, North Yorkshire, England.

Thus *in situ* learning can shape knowing publics through embodied practice and sensory experience, using not only sight but also hearing, smell and touch (e.g. Macnaghten 2003; Macnaghten and Urry 1998; Waitt and Cook 2007), through deliberate environmental education programmes such as those run in national parks. But learning may also be a by-product of practices aimed not at learning but enjoying, such as tourist encounters with unfamiliar, dramatic and exciting environments (e.g. Cloke and Perkins 1998; Urry 1995), as we shall see in Chapter 5. While *ex situ* knowledge-practices of environmental learning are often highly formalised, clearly structured and based in educational institutions, *in situ* practices are often much more informal, entertainment-focussed, diffuse and do not result in qualifications or measurement of learning in the form of grades.

Like any conceptual division, this can be oversimplified, as some environmental knowledge-practices that publics enjoy are not easily categorisable into *in situ* or *ex situ*. Table 2.1 showed how many students chose courses in formal higher education settings that may incorporate environmental learning explicitly in their curricula, offering learning in classrooms, libraries and online 'virtual learning environments', but also in field trips outdoors. In addition, students may also encounter environmental policies and management as they walk about campus or live in dormitories managed by their college or university, because many higher education institutions have sought to improve the environmental management of their own lands and facilities in order to gain higher rankings on 'green' league tables such as that run by People and

Planet in England (https://peopleandplanet.org/university-league), but also to avoid public criticism and to reduce running costs. Students might also participate in conservation events run by student groups or campaigns against environmental damage led by students, such as students from the University of Manchester protesting against BP's sponsorship of Manchester Museum (see www.artnotoil.org.uk/blog/students-protest-against-bp-sponsorship-manchester-museum-exhibition-20411#sthash. Raqtpskx.dpuf) or geography and biology students at my own university picking up litter on the edge of an estuary at the weekend, shown in Figure 2.2.

As well as formal and embodied learning, learning may be shaped through engaging with environments through radio, television, websites and other media. Large audiences now watch environmental science and natural history programmes on television, with over five million people regularly watching the BBC's weekly *Countryfile* programme in the UK, putting it in the top ten most-watched programmes, and over half a million people in the UK regularly watch the *Discovery* channel. Books about natural history such as MacFarlane's *Mountains of the Mind*, *The Old Ways* and *The Wild Places* and Deakins's *Waterlog* and *Wildwood* have also sold well in recent years. Such *ex situ* learning about environments may take place in comfy living rooms or whiling away the time waiting at airports or on train journeys, being not only highly informal but also carried out for fun, suggesting widespread public interest in informal learning as entertainment (or what has been termed infotainment). Such practices also blur the boundary between knowing and consuming (see Chapter 5

FIGURE 2.2 Students from the University of Hull picking up litter on Humber estuary with the Yorkshire Wildlife Trust.

and Michael 1998) and between knowing and enjoying, again emphasising the necessarily artificial nature of the divides between chapters in this book.

Ordinary people often report high interest in science, hence the prevalence of science stories in news and entertainment media: as I type this, a story about water being found on Mars is on the front page of my newspaper, for instance. But ordinary people also say in surveys that they have only low knowledge about science (e.g. Durant et al. 1989), again making it difficult to measure the resulting scientific or environmental 'literacy'. In 2010, only 11 per cent of people surveyed across Europe (17 per cent of those surveyed in the UK) felt "well informed" about scientific discoveries in general, although many more (30 per cent across Europe and 43 per cent in the UK) said that they were "interested" in scientific discoveries. When TNS (2011 Eurobarometer 365) asked respondents to name three main sources of *information* about the environment, television news came top (73 per cent), with newspapers far behind in second place (41 per cent), which suggests that mass media editors have a great influence on how publics see and understand environmental issues, through both how they select and how they tell environmental stories. But when respondents were asked which sources they most *trust* regarding environmental issues, scientists topped the poll (40 per cent of respondents), followed by environmental groups like Greenpeace and WWF (37 per cent), with television far down the ranking (29 per cent) and government, private companies and trade unions at the bottom with less than 10 per cent each. So again, practices of learning do not necessarily map onto practices of trusting or judging information nor of following its prescriptions.

Importantly, unlike formal higher education, casual learning *in situ* in particular may take place without being recognised as such, bundled as it is within and through other practices. People may regularly experience a particular environment and build up a store of knowledge and feelings about it, so that the knowledge gained cannot be easily (a) validated by scientific means or (b) separated from the bodily and emotional experience of going to that environment as part of their personal life, which may also make articulating that knowledge in disembodied forms (such as responses to survey questions) difficult.

Let me give a more detailed example to show what I mean. People who go fishing as a hobby engage with the environment, particularly the water environment, and their recreational practices involve not only *doing* fishing but also *learning about* fishing in different ways, that is, through different environmental knowledge-practices both *in situ* and *ex situ*. *In situ* examples include looking at the river's surface to interpret the visible pattern of eddies and swirls as indications not only of movements in the water made by fish, but also of the shape of the underlying river bed, even though both fish and river bed remain hidden from human sight. Anglers take photographs of their catch as trophies, but also of the environments as records, supporting visual learning *ex situ* as they compare photos against each other over time and against maps to build up layers of environmental knowledge across seasons and watersheds, building folk memories of fluidity, of the suspended-sediment colour of their favourite river at peak flood, of the riverbed substrates exposed during droughty periods. They may also record water temperature over time, creating their own personalised databases of past climate change, which they then use to decide where, when and how to fish in future.

But learning is also embodied through the embodied experience of casting, baiting, reeling in, with the constant feedback of the water's pull on the line through the angler's fingers, the drag of the sediments on the river bed, the sinking of the bait, the coolness of the water coursing around the outside of their wading boots. This emphasises that environmental knowing is also about environmental doing (e.g. Law and Mol 2002; see Chapter 1), that practices both support learning and are learning about environmental conditions and changes. And such practices include not only the angler and the water, but other devices beyond the fishing line, because practices are complex assemblages of diverse elements and dimensions (e.g. Shove and Pantzar 2005; Truninger 2011; see Chapter 1). Thermometers are used by anglers to record changing temperatures and help them choose which gear and bait to use and even which fish species to target; plumblines will check for depth in the target zone, as the lead weight descends through the water column; polarised spectacles reduce glare on a bright day, allowing anglers to see beneath the water surface a little more successfully.

Knowledge-practices are also shared, as we saw in Chapter 1, through angling magazines and online forums full of tips as to the best places to fish, the best gear and bait to use in particular weather conditions, how target fish species live and respond to their own environment. Reading and writing are further environmental learning practices that here support not formal education attainment, but socialising, competition and fun (as we will also see in Chapter 5). Pooling environmental experiences, anglers not only try to fish better but also try to make sense of environmental changes, such as whether numbers of fish species are declining, how climate is changing, whether predators are increasing and who is responsible for pollution events.

And this learning is not merely for the individual angler or their club or online community. Anglers, like canoeists, farmers and other environmental publics, are often invited to attend environmental management forums in order to tap into their specialised knowledge about water environments and to use this to supplement more official, expert, technocratic knowledge for environmental decision-making. For example, the Environment Agency for England and Wales invited anglers onto its Regional Fisheries, Ecology and Recreation Advisory Committees (RFERACs), recognising their input as valuable precisely because of the environmental practices in which they regularly participate. Informal environmental knowing by these recreational publics thus feeds directly into formal systems of regulation.

And anglers also produce environmental knowledge for the Agency, in the form of catch returns that they submit, either individually or collectively through a club, recording where, when and how many fish (by number or weight) were caught. This data subsidy becomes part of official monitoring and regulation as, alongside other data from automatic fish counters and Agency staff surveys, the catch returns are used to evaluate fish stocks and thus to shape ongoing environmental policy to manage fisheries and water ecologies (Eden 2012). Instead of simply being consumers of knowledge from formally trained or qualified experts, the specialised environmental public who go recreational angling are also knowledge producers (and we shall return to this important distinction later).

Finally, it is worth noting that many environmental professionals start out as environmental recreationists. Anglers become Environment Agency staff, birdwatchers join the Royal Society for the Protection of Birds, gardeners join the National Trust – in each case their environmental learning through leisure becomes also part of environmental management, often supplemented through more formal learning qualifications, such as a fisheries management degree. In a study about river management (Eden and Bear 2012), five of the eight Environment Agency professionals that our research team interviewed were recreational anglers and another two used to be but had now (in their words) 'lapsed'. In such cases, learning, enjoying and eventually working practices became bundled together, influencing each other and enabling publics to perform their multiple identities, whether as recreational users of the environment or environmental workers (see Chapter 8).

Unknowing publics and the deficit model of public understanding

So far, I have outlined a variety of knowledge-practices through which environmental publics know, learn about and experience the environment, enacted through formal and informal, embodied and virtual ways. But often many of these ways are ignored or devalued, so that environmental publics are constructed not as knowing, but as unknowing, as poorly informed or even misinformed about environmental processes, how they are changing and human roles in those changes. So we also need to consider how publics are imagined and enacted as *unknowing*.

Work in the public understanding of science (PUS) is important here. For example, Irwin and Wynne (1996, p. 215) argue that 'the public' is generally constructed by policy makers and other elites as being:

a. "an aggregate of atomised individuals" without their own cultural ideas or connections to each other by which they might share ideas;
b. ignorant because of a lack of intellect or expertise (rather than actively choosing not to seek knowledge);
c. keen on maximising control over processes (such as environmental change);
d. keen on certainty and naively wanting to avoid all risks (also Wynne 2005);
e. homogeneous in access to knowledge.

These assumptions are quite different from the picture of publics I have painted so far in this chapter, but they underpin a dichotomy between 'lay' (public) and 'expert' (scientific) knowledges that continues to influence environmental debates. This dichotomy distinguishes 'expert' scientific knowledge about the environment that comes from formal education or professional training from 'lay' knowledge about the environment that comes through ordinary people engaging with the environment through their everyday lives, such as walking through a park, swimming in a lake or going fishing. It has also been related to 'the distance effect' (Collins 1987), where (social) distance from science lends enchantment and reduces uncertainty. Here, 'distance' is not about mileage but about practices happening in different places, e.g. scientific knowledge-practices happening in laboratories or research conferences, so that they become invisible to those outside those places, that is, non-scientists. In other words, the argument runs, because publics are remote from scientific activities, they are unaware and intolerant of uncertainties and disputes about issues like climate change and GM, as well as lacking understanding or even misunderstanding the basic scientific principles involved.

When the 'lay/expert dichotomy' arises in environmental debates, it is often accompanied by the assumptions of a 'deficit model' of public understanding of science.[3] The deficit model assumes that the public's lack of understanding about an issue (a) is due to a deficit of (correct) information and (b) can and should be remedied by providing information from experts, to generate more understanding and also more trust, legitimation and support for how the state (or other institutions) are choosing to deal with the issue in question. Surveys (e.g. Lorenzoni et al. 2007, p. 454) report that ordinary people are often uncertain about climate change, for example, confusing it with ozone depletion or recycling or thinking that the scientific consensus about climate change is far less strong than it is, leading to the argument that more information should be provided to correct such misperceptions by the public.

However, social scientists have shown that things are not as simple as the deficit model assumes. Knowing about environmental change is bundled up with other sorts of knowing, such as about how pollution or flooding has been handled by state agencies in the past, about the competence of local authorities in planning processes, or about failures of scientific governance to anticipate and manage risks from new technologies. In other words, public knowledge about environmental risks is also an expression of public views about risk and the organisations that manage risks (or attempt to manage risks) and cannot be separated from them.

> What scientists interpret as a naive and impracticable public expectation of a zero-risk environment can thus be seen instead as an expression of zero trust in institutions which claim to be able to manage large-scale risks throughout society.
>
> *Irwin and Wynne 1996, p. 218*

So public perceptions may differ from 'expert' perceptions not merely because the public are unknowing, but for other reasons. Maybe the public have different personal priorities and moral values from those used in an expert assessment, or they draw on a contextual history of past mismanagement by state or commercial organisations when imagining what will happen in the future; perhaps they see flaws and uncertainties in 'expert' assessments that cause them to distrust the outcomes of those assessments, or they see 'experts' as outsiders with hidden and unwanted agendas – these can all influence how people understand an environmental issue.

One famous example is the study by Brian Wynne (1992, 1996) of how the impact of fallout from the Chernobyl nuclear accident on the landscape and ecology of northern England was understood by 'lay' farmers compared to 'expert' scientists. He argued that the scientific consensus about the fallout failed to appreciate the farmers' lay knowledge, which was 'black-boxed' or suppressed in the official analyses. For their part, the farmers distrusted the scientific analysis, feeling manipulated for scientific data collection rather than benefiting from it.

Importantly, rather than arguing that one side was right and the other wrong, Wynne (1996, p. 38) argued that *both* sides were equally "socially grounded, conditional and value-laden", but that the scientific side was less reflexively aware of this. So instead of seeing science as 'right' and the farmers as ignorant or misinformed, Wynne's argument dismissed the deficit model and instead highlighted how knowing is contextual and also built upon unequal power differentials, in this case in terms of evaluating the safety of animals and landscapes.[4] Similarly, Irwin (1995 p. 114–115) argued that farm workers disputed the 'expert' view of pesticide safety because they were more familiar with how pesticide sprays are used in practice, that is, their knowing came from doing, and that doing did not strictly follow the safety guidance as 'experts' assumed.

In both cases, the 'expert' recommendation or consensus was based on a modernist notion of control, assumed uniformity of conditions and standardised practices and universality of knowledge. By contrast, the 'lay' view was based on a reflexivity and an appreciation of the messiness, local differentiation and heterogeneity of social and environmental contexts. In another study of farmers, this time in connection with wetland protection, Burgess et al. (2000, 127) found that conservationists in government agencies sought to control and manipulate nature, especially through the use of science, again in a typically modernist conceptualisation, but farmers perceived the environment to be dynamic and capable of balancing itself, if left alone.

What all this shows is that, rather than a lay/expert dichotomy with two realms opposing each other and defined by different levels of environmental knowing, we need to recognise a more complex pattern of knowing and unknowing publics and of publics knowing different things about different issues in different places, and to acknowledge that people can move across this complex archipelago of different beliefs and associations, as they learn and change.

> The assumption of homogeneity of perspectives in the expert/lay dichotomy obscures significant variation in the perceptions and preferences of both. It also obscures the fact that real people are not consistently experts or lay people. There are no universal experts and in the civic arena, even the most modest lay person has some relevant expertise.
>
> *Thompson and Rayner 1998, p. 336*

And of course the same is true of 'experts', who also differ; for example, not all subscribe unthinkingly to the 'deficit model' and different sorts of 'experts' imagine the public in different ways according to context (e.g. Blok et al. 2008). So we need to recognise how the officials professionally engaging with the public are also differentiated and reflexively perform their roles, influencing and co-producing themselves and their publics through environmental debate and knowledge-practices.

How people understand and predict environmental change is thus not some disembodied, objectively framed calculation about the future, but is part and parcel of how they make sense of their social worlds more generally, including how they make sense of and judge politics, policy and science; environmental knowledge is thus necessarily and pervasively also social knowledge. Wagner (2007) thus argued that "vernacular science" (including metaphors, imagery and scientifically incorrect knowledge) is an important stage of the way that people move from a position of no interest about an issue to being more informed: one function of 'vernacular science' is to be socially useful, enabling people to take part in social, informal conversation about issues in the news, for example. So new information about environmental issues is not simply accepted by people in isolation; rather, it is sifted and judged according to the information and beliefs people already have about the environment, but also about the trustworthiness and purposes of scientists, policy makers and other elites (Thompson and Rayner 1998). The knowing practices of environmental publics are not performed in a vacuum or on a blank slate, but are part of an ongoing social process that has both context and history.

Social scientists have been quite successful at pointing out the problems with the deficit model since the 1990s, and challenging policy makers to open up technocratically framed debates to public input. Today, many governments and other powerful actors (such as companies) officially endorse a model of public engagement over environmental decisions based on dialogue and two-way learning between experts, policy makers and the public, suggesting a move away from the deficit model at least *in principle*. But Irwin (2006) argued that public engagement exercises still tend to implicitly frame debates in ways that work against dialogue *in practice* – a case of talking the talk of 'lay' input, but not walking the walk. In other words, although the official line is that publics should be involved in environmental debates in a two–way exchange of views with official, expert or elite groups, in practice public engagement is still designed and structured on the assumption that information will be given to the (unknowing) public and they can only respond to it, rather than shape it.

Place and scale in knowledge-practices

Let us move on to consider the geographies of these knowledge-practices. Place is an important contextual factor in the arguments made so far in this chapter, yet is often neglected by work in the public understanding of science (PUS) that is dominated by sociologists.

Geographical context influences environmental publics in multiple ways. For example, the media coverage of environmental issues varies from country to country, as well as from national outlets such as national newspapers and broadcast television, to local ones. Poberezhskaya (2015) compares the relative silence on climate change in national media coverage in Russia with the far greater volumes of stories in national media in the UK and USA, estimating that the USA has approximately five times the volume of climate change stories than in Russia, and the UK approximately ten times. Environmental publics in each country are thus exposed to and find different amounts of media coverage available through very different styles of story-telling, from reporting new science to criticising the inaction of politicians.

But there are other contextual factors beyond the geographical location of environmental publics. The *in situ/ex situ* distinction that I discussed earlier highlighted where learning takes place and how that location shapes the purpose and mode of knowing that may result. For recreational anglers, embodied *in situ* learning shapes practice, although plenty of time may also be spent *ex situ* reading fishing blogs or magazines for fun, and sometimes *ex situ* learning may occur in the same place and time as *in situ*, now that fishing ponds increasingly providing Wi-Fi internet access so that anglers can blog and surf as well as cast and catch. So the geographies of where learning happens both afford particular sensory and intellectual engagements, while limiting others.

And there is also the *scale* of learning practices. Scale is often thought about in hierarchical terms, with the higher scales of knowledge more valued in environmental debates. For example, science is classically assumed to be seeking universal laws of reality, laws that apply regardless of particular conditions and geographical contexts, such as laws of thermodynamics, laws that are therefore universal in scale. But the assumption that knowledge should (and can) be universal is particularly problematic for *environmental* science where local conditions are highly variable and influential, so that environmental science is often argued to be messier, less amenable to universal laws than other sciences, such as physics.

This ideal of universal knowledge also underpins some of the lay/expert dichotomy that I discussed earlier. This is because scientific knowledge, even where it is not strictly 'universal', is still assumed to be generalisable and involve big-pattern thinking, whereas 'lay' knowledge is assumed to be much more small-scale or piecemeal in its spatial coverage. Laypeople are assumed to know more about (and therefore preferentially value) familiar environments, that is, the local rather than the global, the visible rather than the remote, and such "knowledges for doing" tend to be "highly practical, case-specific and instrumental in orientation" (Irwin 1995, p. 133). Hence, science is often associated with higher scales of environmental knowledge and public views with much lower scales. For example, Lorenzoni et al. (2007) reported that many ordinary people that they surveyed focussed on environmental concerns that were more local, more tangible and more immediate in terms of timing than global concerns like climate change.

This has consequences for how knowledge is valued, because 'expert' criticisms of 'lay' knowledge assume that lay knowledge is locally bound and not readily transferable to other scales or places (also Irwin 1995, p. 117). This problematic hierarchy of knowing is built therefore not upon *what* is known but *how* it is known – formal qualifications, professional positions and other credentials are used more frequently to legitimate environmental knowledgeability than are recreational experiences, for instance. This means that expert knowing may be valued above non-expert knowing and thus be more able to influence decision-making by elites, such as governments.

The opposite side of the coin is that in some cases there has been a push to (re)value local lay knowledge in different ways. Researchers in development studies in particular have attempted to shift how we regard the indigenous knowledges of lay people in developing countries from being dismissed to being valued, even "privileged" in development debates (e.g. Agrawal 2002, p. 287). In environmental terms, such knowledges may be especially detailed about local fauna and flora, such as species which may not (yet) be recognised, named and studied by formal systems of western science, thus offering value in quite practical ways. This does bring its own problems: where such indigenous knowledge is valuable, it may be exploited by pharmaceutical companies looking for new and profitable drugs, thus "harvesting" (Agrawal 2002, p. 288) the knowledge as resource, stripping away its cultural significance through abstraction and 'scientisation' (also Wynne 2005) and undermining intellectual property rights and possibly later income from royalties for indigenous people. And this may be done by companies based outside the countries being harvested, that is, they are geographically elsewhere, footloose rather than embedded in that context. But in terms of conceptualising environmental publics as knowing, such arguments still clearly accept and value local knowledge, despite its lack of scientific universality.

But in western societies, lay knowledge is often regarded far more negatively, with local knowing frequently associated with obstructive NIMBY (Not In My Backyard) protests against developments such as incinerators, roads and airport runways. Academics now largely see NIMBYism as an outdated and oversimplistic way of explaining opposition by local publics (e.g. Aitken 2009; McClymont and O'Hare 2008; Wolsink 2006), because it is often used to imply that local publics are ignorant or misinformed about the impacts of a proposed facility, whereas they are often highly informed from both local experience and also because they pay more attention to media coverage of issues local to them (Devine-Wright 2009). Despite this, the NIMBY label is still frequently used to belittle local concerns, asymmetrically constructing local opponents of a new facility as biased and unknowing publics, motivated only by self-interest, compared to proponents, and also deposing local knowledge to the lowly, least-valued rung of the scalar hierarchy.

So here are two different constructs: a knowing public in the form of indigenous peoples in developing countries who may be employed as 'experts' to identify valuable local flora and fauna for visiting scientists or bio-prospectors from elsewhere, and an unknowing public in the form of ordinary people in more developed countries who do not understand environmental processes and instead need 'experts' to judge for them whether local developments will damage valued environments. Different constructs of un/knowing publics may thus carry different meanings and levels of value in different places, emphasising that situated, embedded and contextual 'lay' knowledges are intimately geographical as well as social, affected not only by the place in which knowing happens but also more generally by assumed scalar hierarchies of worth. And these imaginaries, these meanings ascribed to environmental publics, also shape other practices by which those publics are spoken to, involved (see Chapter 3) or opposed (see Chapter 6) by other elites, and through which environmental publics are continually re-imagined and re-made (see Chapter 1).

Power

To sum up so far, formal qualifications in environmental science are rarely found amongst members of the public, meaning that environmental publics are commonly imagined (especially by policy makers and scientists) as *unknowing* about the environment, or at least only knowing about *local* environments, a knowing which can often be dismissed by decision makers as biased or unhelpful (e.g. Owens 2000; Ungar 2000; Wynne 2005) and less worthy because of its local scale.

These imaginaries have profound consequences for power. They can render publics as merely passive witnesses to science or environmental decision-making, not active participants in knowledge production and application. If the risks and effects of environmental decisions are assumed to be too technical for laypeople to fully appreciate, then only the 'experts' will be assumed to be qualified enough to decide. Being constructed as ignorant, the public "can therefore be legitimately excluded from influence", limiting their power in traditional politics and decision-making, so that "'citizenship' only begins when 'expertise' has set the environmental agenda" (Irwin 1995, p. 79), that is, the public only enter debates about environmental issues after science has shaped the remits and possible solutions of those debates.

And this image of 'the public' as unknowing can also influence how publics imagine themselves, which might produce self-exclusion, deference and derogation of decision-making powers to those perceived as experts and have knock-on effects in terms of how (and indeed whether) publics choose to contribute to engagement exercises (see Chapters 3 and 6) and other democratic processes (see Chapter 7).

Knowing too much: Differentiating publics by power

That said, the 'lay public' are not all the same – they are not "an undifferentiated blob" (Thompson and Rayner 1998, p. 274) but highly diverse in how they think about science, the environment and their own actions, as well as how much knowledge they have and how confident they are in using it in environmental debates.

For example, during the UK 'GM Nation?' debate in 2003 about how the UK should manage genetically modified crops, the extensive public participation process was criticised (especially by scientific experts) for being dominated by people who already had views on GM crops and by organised 'green groups' (Horlick-Jones et al. 2007; Irwin 2006). In this sense, the ideal of unknowing, open-minded, unbiased public input was felt to be swayed by activists with pre-existing opinions, what are sometimes called 'public interest groups', but what I called 'specialised publics' (Eden and Bear 2012) and Michael (2009) called 'publics-in-particular'. In the 'GM Nation?' case, such groups were criticised for not being 'innocent' or 'true' representatives of the general public, because they were (no longer) open to persuasion by experts in line with the deficit model – they had failed to fit the assumption about the general public being unknowledgeable and therefore persuadable, so they had less 'deficit' than was usually assumed in the 'deficit model' discussed earlier.

So power reflects implicit views about who should be listened to and who should have sway in environmental decisions. In the 'GM Nation?' case, activists or others with pre-existing interests were criticised for not being legitimate representatives of this assumed 'general public', but in some way skewed or biased. Similarly, when (especially local) publics oppose new developments such as wind farms, incinerators, facilities for homeless people or airport runways, their views can be denigrated as NIMBYism on the basis of parochial, embedded self-interest, as I noted earlier.

We can therefore see that, as well as the expert/lay dichotomy, there is another dichotomy constructed that divides 'the public' into 'good' and 'bad' publics (Eden and Bear 2012; Michael 2009). 'Good' environmental publics are those imagined to be motivated by altruism and civic interest, whereas 'bad' environmental publics are those imagined to be motivated by NIMBYism and self-interest, often also regarded as 'specialised publics' because they are more knowledgeable and often more active in campaigning about 'their' issues.

This knowledgeability can be seen positively and exploited when state agencies co-opt such publics into policy communities to tap into their knowledge and reduce challenges to policy. Specialised publics may also establish nongovernmental organisations to carry forward their interests or work through more informal networks of active citizens, as noted during the 'GM Nation?' debate. But often knowledgeable publics are seen negatively, cast as specialised and thus unrepresentative, no longer 'the general public' but biased because of their narrowly defined interests, whether these are sociodemographic, geographical or hobby-related, and perhaps misinformed, rather than under-informed, about environmental issues.

As social scientists, we need to seek out these differentials, to pull apart the often-assumed imaginary of 'the public' as singular or homogeneous. Irwin and Michael (2003, p. 85) emphasise heterogeneity in scientific citizenship, arguing that rather than thinking of 'the public' or assuming a lay/expert dichotomy, we need to recognise numerous 'ethno-epistemic assemblages', that is, coalitions or movements that shift and reconfigure expertise in more fluid, hybrid ways, that mix science and society together so that, rather than being produced and understood solely within pockets labelled as 'science', knowledge is distributed in more complicated ways through society. And, as I have emphasised in this chapter, knowing is also geographically performed, complicated not only by the diverse places in which learning happens but also by the scaling of value accorded to both environmental knowing and the publics that express and are enacted through such knowing.

Power and the environmental amateur

There is another angle to this argument about power relations and knowledgeable publics, which is where highly specialised publics already have or can collect environmental knowledge that is useful to the state, research institutions and environmental organisations. Such environmental publics know and also may produce environmental knowledge and, where this knowledge is valuable, those publics become valued by consequence. For example, state regulators increasingly use lay volunteers to gather data about environmental change as part of official monitoring systems, not least as budgets in the state sector are cut because of economic recession. Sometimes the state can tap into existing networks of knowledgeable publics who collect environmental data as part of their hobby, data that state organisations can then request from them; sometimes, publics are specifically recruited and trained to collect new data for the state's purposes and to its specifications.

Amateur naturalists exemplify this developing relationship of knowledge co-production between environmental publics and the state. An individual's passion for finding and recording environmental phenomena, whether these are mosses, birds or butterflies, can provide information for free as a useful subsidy for state agencies in a time of reduced budgets. Enrolling enthusiastic amateurs as (paid or unpaid) volunteers can therefore extend the amount and scope of environmental fieldwork that conservation and environmental science can do. For example, amateur naturalists were enrolled by English Nature and the Natural History Museum to gather biodiversity data for them (Ellis and Waterton 2005) and data collected mainly by volunteers for the British Trust for Ornithology about British farmland birds has been used not only to indicate likely declines in species and thus biodiversity over the decades (Greenwood 2003), but also to measure sustainable development generally. In the USA, the Cornell Lab for Ornithology and the National Audubon Society have run the Great Backyard Bird Count since 1998 as part of a wider programme of 'citizen science' (Bonney et al. 2009), and the UK's Royal Society for the Protection of Birds has run Garden Birdwatch since 1979, with up to 280,000 people reporting birds sighted in their gardens on a specific day. Recreational anglers also contribute their catch data to state agencies in similar ways, as I noted earlier.

Do such practices blur the boundary between professionals and amateurs (Star and Griesemer 1989; Kuklick and Kohler 1996)? To put it another way, do these environmental publics become less 'public' and more 'scientific' through co-producing or contributing knowledge? Usually not, because professional organisations do 'boundary work' (Gieryn 1995; Kinchy and Kleinman 2003) to maintain the spatial demarcation of expertise: the 'amateurs' are usually out in the field, whether this is defined as their garden or open countryside, and not in the laboratories or the offices of paid 'professionals', that is, they are not in what we might think of as the centres of calculation (Latour 1987). Ellis and Waterton (2005, p. 686) showed how 'novice' amateur naturalists had an 'apprenticeship' to an acknowledged 'expert', who verified their plant classification either *in situ* while doing fieldwork together or *ex situ* when a sample was sent by the 'novice' to the 'expert', echoing the geographical separation discussed above. Even highly specialised publics with explicitly useful skills and knowledge are therefore explicitly separated from the professional, usually formally qualified cadre of environmental managers. So even seeing such 'lay' (unpaid, unqualified) people as more knowledgeable than usual does not mean granting them power in environmental decision-making; their involvement is still managed by the state and its officials, who may often check the amateur's credentials and their practices to validate their knowledge, rendering them subordinate to the more knowing 'expert' employed by the state.

Some hope that amateur enrolment and involving publics in producing knowledge (rather than solely consuming it) will improve the relationship between science and society, not only for environmental matters but also for health, energy production and new technology. Gibbons et al. (1994) term this "Mode 2 knowledge production", which, as a supplement to the classical, homogeneous, disciplinary-focussed science of 'Mode 1', involves a broader range of differentiated knowledge producers in generating knowledge which is contextualised, useful and socially accountable. Irwin (1995) refers to 'citizen science', taking this as the title for his book and applying it to a wide range of initiatives and issues, from GM food to air pollution. A variety of other terms are also used by researchers, such as 'volunteer monitoring', 'community-based monitoring', 'participatory monitoring' and public participation in scientific research (Jalbert and Kinchy 2015; see Shirk et al. 2012 for comparisons and a typology).

In the environmental case, this changing science–society relationship is necessary and timely because science increasingly is seen to be unable to handle the sort of complex, transdisciplinary problems of understanding and managing environmental change. But this also offers an opportunity for science and expertise to be opened up and democratised (Jasanoff 2003), for a 'greener' form of science to be open and self-critical (Wynne and Mayer 1993), for "participatory expertise" to become a conscious part of the pluralisation of knowledge (Fischer 1990) and for an "extended peer community" to be involved (Functowicz and Ravetz 1990). Such arguments all suggest a sharing of the power that knowledge (or at least the assumption that one has knowledge) may claim.

Does this power-sharing work in practice though? Ottinger (2010) showed how residents of Louisiana took their own air quality samples using 'bucket' kits to demonstrate high levels of pollution by local industry that could damage their health. In this case of 'citizen science', lay people were grudgingly accepted by elites as producers of environmental knowledge because they used standard analytical techniques to measure air pollution, but the impact of their results was limited because they used non-standard techniques to collect their samples of air. This was sometimes effective in prompting regulators to check out particularly high readings to see if there had been unreported accidents, but was ineffective in changing legislative standards for air quality more generally. Here, the local publics were admitted to be knowing and to have data that the official regulators did not have, but their results were seen as less reliable than those of professional regulators, who used standardised methods.

As noted above, state agencies sometimes draw on lay knowledge as a form of subsidy where they have insufficient resources to study environmental change themselves. The Environment Agency of England and Wales uses recreational anglers to gather data about fish populations, requiring anglers to complete log-books about their individual yearly catch of 'game' fish species (chiefly salmon and sea trout) and angling club secretaries to complete 'catch returns' about the collective catch of 'coarse' fish species (e.g. chub, roach, bream, dace) by their members during day matches. These data feed directly into national evaluations of fish stocks and then into international evaluations of North Atlantic salmon stocks, thus influencing decisions about how best to manage ocean fisheries.

If not exactly professionalising the amateur, do such knowledge-practices at least suggest a new role, a new form of citizenship, for environmental publics? Ellis and Waterton (2004) suggested that it might, although this depends very much on how the state and the communities of amateur naturalists see each other. If amateurs are used merely to get good environmental data out them, what Ellis and Waterton (2004) referred to as an 'extractive' strategy, then perhaps not, because building public capacity and better state-public relationships through coproducing environmental knowledge needs to be fully contextualised and nurtured through two-way communication and participation with volunteers. For example, the British Trust for Ornithology aimed to give their volunteers "ownership of the work from the start" and "value and cherish them not just as fieldworkers but as a network of well-informed people who help take messages out to the wider community" (Greenwood 2003, p. 228). Such projects aim to sustain volunteer engagement through bonds of information. For example, Cornell Lab for Ornithology's 'eBird' website was redesigned so that participants who uploaded data could compare their data with data uploaded by other participants, and afterwards "the number of individuals submitting data nearly tripled" (Bonney et al. 2009, p. 981), by implication because people were finding the process more interesting, interactive and informative. More recently, smartphone apps have been developed to log sightings of key species as part of research projects, such as the 'Leafwatch' app for sightings of the horse-chestnut leaf-mining moth, *Cameraria ohridella*, to monitor its spread across England (see www.conkertreescience.org.uk/leafwatch-app; Pocock et al. 2014).

Amateurs may help to produce knowledge for altruistic reasons, to further environmental understanding generally and in the public interest, but they mainly do it for their own satisfaction and fun (Ellis and Waterton 2005; Lawrence 2006). Money may also help as an incentive, for example where angling clubs have been offered payment to submit catch returns (fish counts by weight and species) from their matches and anglers returning completed individual log books have been offered entry to a prize draw by the Environment Agency (Eden 2012). So, knowledge-producing publics may be paid for their efforts, but still be regarded as amateurs.

Self-learning, activism and lay expertise

Finally, we should consider the knowledge-practices in activism, where gaining environmental knowledge is used for self-empowerment and campaigning (see also Chapter 6). Laypeople may be prompted by threats to their health or to a local environment to self-educate and become (more) environmentally knowing. Getting involved in (or even starting up) environmental campaigns is both a motivation for and a consequence of environmental learning: people can become very knowledgeable about, for example, waste management techniques because they want to be able to judge the potential impact of plans to build an incinerator in their local area, or people may learn about genetic modification of food through internet websites and go on to seek out political means to express their concerns. Again, this emphasises how knowledge-practices intimately entangle both knowing and doing and also how important geographies of knowing are in performing environmental publics.

There are a few studies of how people change from being the idealised 'unknowing' lay person to a highly knowledgeable, often highly active member of a campaigning group. Epstein (1995) studied what he considered to be the 'unique' case of AIDS campaigners, which showed how people that suffer from medical conditions often learn a great deal about those conditions in order both to engage in complex debates with the medical profession about treatment options and to participate in trials. The intensive self-education in such cases can result in "the mass conversion of disease 'victims' into activist-experts" (Epstein 1995, p. 414), who were listened to more by elites in designing medical trials. Similarly, 'citizen science' or other forms of public participation in scientific research can also benefit participants by developing their understanding, data collection skills and scientific or environmental literacy (Shirk et al. 2012).

> Participants are empowered by the practice of learning about how their ecosystems work – and their fragility. Participants also educate one another on how to navigate complex legal systems. Perhaps most importantly, concerned citizens realize they can play a part in pressuring government agencies to take seriously their responsibilities in protecting residents' well-being.
>
> *Jalbert and Kinchy 2015, p. 14*

Public participation in knowledge production can also change publics by enhancing their sense of stewardship, expertise, confidence in their own knowledge and even greater personal 'ownership' of projects, potentially strengthening people's relationships not only with particular environments and/or species being monitored, but also socially with other people involved in those monitoring networks (Shirk et al. 2012).

So, as Irwin (1995) pointed out, "we are all part of social experiments about the environmental consequences of technological development", whether these are the environmental consequences of nuclear power stations releasing radioactivity or of GM crops being planted in fields: the public are now in the lab because these environmental consequences are happening in the real world and affect us. And the corollary of that is that our input to decisions about how we use and regulate nuclear power, GM crops and many other things is important and necessary. But that still does not guarantee a role or any power to influence very technically framed environmental decisions.

Summing up

This chapter has challenged common assumptions that environmental publics are unknowing or, even where they do produce or demonstrate environmental knowledge that is relevant to debates, that their knowledge can only ever be local or non-expert and thus lack value for decision-making. These assumptions reflect how environmental publics are imagined but they also shape how engagement exercises are designed to reach out to and engage these imagined publics. Instead of assuming a lack of knowledge as a hole to be filled by information, we should concentrate on how environmental publics perform and are performed through knowledge-practices, that is, not on what people know but how they know it, how they see themselves as knowing publics and how others see them as un/knowing publics.

I have also emphasised how environmental publics are highly differentiated in the level and types of environmental knowledge that they seek to gain (and distribute). Different sorts of 'specialised publics' make different claims to environmental expertise. Some environmental publics also produce environmental knowledge, not merely consume it, and they do this sometimes for their own satisfaction or purposes, sometimes for other purposes (such as to support state monitoring and policy) and sometimes for both reasons.

A final question for this chapter is whether knowledge changes people's practices. Studies suggest that providing information about an environmental or scientific issue does not produce knowledge about it, nor necessarily engender support for or trust in the way the issue is being managed by official experts or elites; indeed, providing information may increase opposition and uncertainty. For example, Dryzek et al. (2009) analysed various consensus conferences in the USA, Canada, France, Australia, Switzerland and the UK in 1998–2003, and despite deploying different participatory mechanisms for selecting participants, organising conferences and supplying them with information and expertise, they found that the publics involved still tended to become more precautionary about GM food during the process. Publics involved in the 'GM Nation?' exercises seemed to become more ambivalent about GM food as a consequence (Horlick-Jones et al. 2007, p. 104), judging new information in accordance with their pre-existing beliefs, whether positive or negative. This 'confirmatory bias' means that people tend to select information that fits in with their existing views (Poortinga and Pidgeon 2004; also Frewer et al. 2003), so encountering more information often does not change people's views, but may confirm or reinforce them.

Knowledge is also often assumed to be transferred across practices, making learning relevant to all aspects of an individual's life. For example, the UK Eco-Schools programme aims to change not only in-school practices, but at-home practices as well: "to help make every school in the country sustainable and to bring about behaviour change in young people and those connected to them so that good habits learned in schools are followed through into homes and communities" (Eco-Schools England 2015). But knowledge does not necessarily simply transfer from practice to practice, because in spilling across spaces of practice, shifting from place to place, person to person, from formal to informal places of learning, it is not only changed but also enacts the environmental publics that share, extend and deny it in different ways. As we shall see in later chapters, other contexts of everyday life, such as consuming (Chapter 4), enjoying (Chapter 5) and working (Chapter 8) often greatly influence not only what environmental publics do, but also how what they do is understood and valued (or not).

To summarise, knowing publics are both producers of environmental knowledge and are themselves produced by that knowledge. Their power to influence decisions and the spatialisation of their knowledge are both shaped by their knowledge-practices, through the place(s) of learning and the scaling of knowledges. And knowing publics are clearly very diverse: the outdoor passions of rock-climbers, anglers, birdwatchers and gardeners will vary both within and between each group, as well as differ from the indoor environmental pursuits of those watching nature documentaries on television. Knowing more about knowing means that we can no longer talk about 'public opinion' or the views of 'the person in the street', but must always hold before us the great differentiation of environmental publics and the varied practices that enact them.

Notes

1 Based on the statistics for provisional entries issued by Ofqual, the Higher Education monitoring body, in April 2015: www.gov.uk/government/uploads/system/uploads/attachment_data/file/429275/2015-05-22-summer-exam-entries-gcses-level-1-2-certificates-AS-and-a-lev____.pdf
2 Completed for HEFCE, the funding body for higher education in England. See www.hefce.ac.uk/pubs/rereports/Year/2008/sdhefcestrategicreview/Title,92209,en.html
3 For an example of the deficit model being deployed, see Durant et al. 1989; for critical analysis of the model, see Irwin 2006; Irwin and Wynne 1996; Jasanoff and Wynne 1998; Michael 2002; Wynne 1995.
4 Unfortunately, Wynne's famous study is usually cited rather too simplistically as an example of where 'lay' people know 'better' than scientists, a reversal of the deficit model which itself can be problematic.

References

Agrawal, Arun (2002). Indigenous knowledge and the politics of classification. *International Social Science Journal* 54, 173, 287–297.
Aitken, Mhairi (2009). Wind power planning controversies and the construction of "expert" and "lay" knowledges. *Science as Culture* 18, 47–64.

Blok, Anders, Mette Jensen and Pernille Kaltoft (2008). Social identities and risk: expert and lay imaginations on pesticide use. *Public Understanding of Science* 17, 189–209.

Bonney, Rick, Caren B. Cooper, Janis Dickinson, Steve Kelling, Tina Phillips, Kenneth V. Rosenberg and Jennifer Shirk (2009). Citizen science: a developing tool for expanding science knowledge and scientific literacy. *BioScience* 59, 11, 977–984.

Burgess, Jacquelin, Judy Clark and Carolyn M. Harrison (2000). Knowledges in action: an actor network analysis of a wetland agri-environment scheme. *Ecological Economics* 35, 119–132.

Cloke, Paul and Harvey C. Perkins (1998). "Cracking the canyon with the awesome foursome": representations of adventure tourism in New Zealand. *Environment and Planning D: Society and Space* 16, 185–218.

Collins, H.M. (1987). Certainty and the public understanding of science: science on television. *Social Studies of Science* 17, 4, 689–713.

Devine-Wright, Patrick (2009). Rethinking NIMBYism: the role of place attachment and place identity in explaining place-protective action. *Journal of Community & Applied Social Psychology* 19, 426–441.

Durant, John R., Geoffrey A. Evans and Geoffrey P. Thomas (1989). The public understanding of science. *Nature* 340, 6 July, 11–14.

Eco-Schools England (2015). *The Eco-Schools Programme*. www.eco-schools.org.uk/ aboutecoschools/theprogramme

Eden, Sally (2012). Counting fish: performative data, anglers' knowledge-practices and environmental measurement. *Geoforum* 43, 1014–1023.

Eden, Sally and Christopher Bear (2012). The good, the bad, and the hands-on: constructs of public participation, anglers, and lay management of water environments. *Environment and Planning A* 44, 1200–1240.

Ellis, Rebecca and Claire Waterton (2004). Environmental citizenship in the making: the participation of volunteer naturalists in UK biological recording and biodiversity policy. *Science and Public Policy* 31, 2, 95–105.

Ellis, Rebecca and Claire Waterton (2005). Caught between the cartographic and the ethnographic imagination: the whereabouts of amateurs, professionals, and nature in knowing biodiversity. *Environment and Planning D: Society and Space* 23, 673–693.

Epstein, Steven (1995). The construction of lay expertise: AIDS activism and the forging of credibility in the reform of clinical trials. *Science, Technology, & Human Values* 20, 4, 408–437.

Fischer, F. (1990). *Technocracy and the Politics of Expertise*. SAGE, Newbury Park.

Frewer, Lynn J., Joachim Scholderer and Lone Bredahl (2003). Communicating about the risks and benefits of genetically modified foods: the mediating role of trust. *Risk Analysis* 23, 6, 1117–1133.

Functowicz, S.O. and J.R. Ravetz (1990). Global environmental issues and the emergence of second order science. Commission of the European Communities Report EUR 12803 EN, Luxembourg.

Gibbons, Michael, Camille Limoges, Helga Nowotny, Simon Schwartzman, Peter Scott and Martin Trow (1994). *The New Production of Knowledge*. SAGE, London.

Gieryn, Thomas (1995). Boundaries of science. 393–444 in Sheila Jasanoff, Gerald E. Markle, James C. Petersen and Trevor Pinch (edited), *Handbook of Science and Technology Studies*. SAGE, London.

Greenwood, J.J.D. (2003). The monitoring of British breeding birds: a success story for conservation science? *The Science of the Total Environment* 310, 221–230.

Holsman, R.H. (2001). The politics of environmental education. *Journal of Environmental Education* 32, 2, 4–7

Horlick-Jones, Tom, John Walls, Gene Rowe, Nick Pidgeon, Wouter Poortinga, Graham Murdock and Tim O'Riordan (2007). *The GM Debate: risk, politics and public engagement.* Routledge, London.

INCA (2012). www.inca.org.uk/france-appendix-mainstream.html#Appendix_5

INRA (Europe) – ECOSA (2000). Eurobarometer 52.1: the Europeans and biotechnology report. http://ec.europa.eu/public_opinion/archives/ebs/ebs_134_en.pdf

Irwin, Alan (1995). *Citizen Science.* Routledge, London.

Irwin, Alan (2006). The politics of talk: coming to terms with the "new" scientific governance. *Social Studies of Science* 36, 2, 299–320.

Irwin, Alan and Brian Wynne (1996). Conclusions. 213–221 in Alan Irwin and Brian Wynne (edited), *Misunderstanding Science? The public reconstruction of science and technology.* Cambridge University Press, Cambridge.

Irwin, Alan and Mike Michael (2003). *Science, Social Theory and Public Knowledge.* Open University Press, Maidenhead.

Jalbert, Kirk and Abby J. Kinchy (2015). Sense and influence: environmental monitoring tools and the power of citizen science. *Journal of Environmental Policy & Planning,* 18, 3, 379–397.

Jasanoff, Sheila (2003). (No?) accounting for expertise. *Science and Public Policy* 30, 3, 157–162.

Kinchy, Abby J. and Daniel Lee Kleinman (2003). Organizing credibility: discursive and organizational orthodoxy on the borders of ecology and politics. *Social Studies of Science* 33, 6, 869–896.

Kuklick, Henrika and Robert E. Kohler (1996). Introduction. Osiris 2nd Series, 11: Science in the Field, 1–14.

Latour, Bruno (1987). *Science in Action: how to follow scientists and engineers through society.* Open University Press, Milton Keynes.

Law, John and Annemarie Mol (2002). Complexities: an introduction. 1–22 in John Law and Annemarie Mol (edited), *Complexities: social studies of knowledge practices.* Duke University Press, Durham.

Lawrence, Anna (2006). "No personal motive?" Volunteers, biodiversity, and the false dichotomies of participation. *Ethics, Place & Environment* 9, 3, 279–298.

Lorenzoni, Irene, Sophie Nicholson-Cole and Lorraine Whitmarsh (2007). Barriers perceived to engaging with climate change among the UK public and their policy implications. *Global Environmental Change* 17, 445–459.

McClymont, Katie and Paul O'Hare (2008). "We're not NIMBYs!" Contrasting local protest groups with idealised conceptions of sustainable communities. *Local Environment* 13, 4, 321–335.

Macnaghten, Phil (2003). Embodying the environment in everyday life practices. The Sociological Review 51, 1, 63–84.

Macnaghten, Phil and John Urry (1998). *Contested Natures.* SAGE, London.

Michael, Mike (1998). Between citizen and consumer: multiplying the meanings of the "public understanding of science". *Public Understanding of Science* 7, 313–327.

Michael, Mike (2009). Publics performing publics: of PiGs, PiPs and politics. *Public Understanding of Science* 18, 5, 617–631.

Office for National Statistics (2014). 2011 Census analysis: social and economic characteristics by length of residence of migrant populations in England and Wales. Office for National Statistics, London. www.ons.gov.uk/ons/dcp171776_381447.pdf

O'Riordan, Timothy (1981). Environmentalism and education. *Journal of Geography in Higher Education* 5, 1, 3–17.

Ottinger, Gwen (2010). Buckets of resistance: standards and the effectiveness of citizen science. *Science, Technology, & Human Values* 35, 2, 244–270.

Owens S. (2000). Engaging the public: information and deliberation in environmental policy. *Environment & Planning A* 32, 1141–1148.

Poberezhskaya, M. (2015). Media coverage of climate change in Russia: governmental bias and climate silence. *Public Understanding of Science* 24, 1, 96–111.

Pocock, Michael J.O. and Darren M. Evans (2014). The success of the horse-chestnut leaf-miner, *Cameraria ohridella*, in the UK revealed with hypothesis-led citizen science. *PLoS ONE* 9, 1, e86226.

Poortinga, Wouter and Nick F. Pidgeon (2004). Trust, the asymmetry principle, and the role of prior beliefs. *Risk Analysis* 24, 6, 1475–1486.

Shirk, J.L., H.L. Ballard, C.C. Wilderman, T. Phillips, A. Wiggins, R. Jordan, E. McCallie, M. Minarchek, B.V. Lewenstein, M.E. Krasny and R. Bonney (2012). Public participation in scientific research: a framework for deliberate design. *Ecology and Society* 17, 2, 29.

Shove, Elizabeth, and Mika Pantzar (2005). Consumers, producers and practices: understanding the invention and reinvention of Nordic walking. *Journal of Consumer Culture* 5, 1, 43–64.

Star, Susan Leigh and James R. Griesemer (1989). Institutional ecology, "translations" and boundary objects: amateurs and professionals in Berkeley's Museum of Vertebrate Zoology, 1907–39. *Social Studies of Science* 19, 3, 387–420.

TNS (2011). Eurobarometer 365: Attitudes of European citizens towards the environment. TNS Opinion and Social, Brussels.

Thompson, Michael and Steve Rayner (1998). Cultural discourses. 265–343 in Steve Rayner and Elizabeth L Malone (edited), *Human Choice and Climate Change: Volume 1: the societal framework*. Battelle Press, Columbus.

Truninger, Monica (2011). Cooking with Bimby in a moment of recruitment: exploring conventions and practice perspectives. *Journal of Consumer Culture* 11, 37–59.

Ungar, Sheldon (2000). Knowledge, ignorance and the popular culture: climate change versus the ozone hole. *Public Understanding of Science* 9, 297–312.

Urry, John (1995). *Consuming places*. Routledge, London.

Wagner, Wolfgang (2007). Vernacular science knowledge: its role in everyday life communication. *Public Understanding of Science* 16, 7–22.

Waitt, Gordon and Cook, Lauren (2007). Leaving nothing but ripples on the water: performing ecotourism natures. *Social & Cultural Geography* 8, 4, 535–550.

Wals, Arjen EJ and John Blewitt (2010). *Third-wave sustainability in higher education: some (inter) national trends and developments*. 55–74 in *Paula Jones, David Selby and Stephen Sterling (edited), Sustainability Education: perspectives and practice across higher education*. Earthscan, London.

Wolsink, M. (2006). Invalid theory impedes our understanding: a critique on the persistence of the language of NIMBY. *Transactions of the Institute of British Geographers* 85, 85–91.

Whitmarsh, Lorraine (2009). What's in a name? Commonalities and differences in public understanding of "climate change" and "global warming". *Public Understanding of Science* 18, 401–420.

Wynne, Brian (1992). Misunderstood misunderstanding: social identities and public uptake of science. *Public Understanding of Science* 1, 3, 281–304.

Wynne, Brian (1995). Public understanding of science. 361-388 in Sheila Janasoff, Gerald E. Markle, James C. Petersen and Trevor Pinch (edited). *Handbook of Science and Technology Studies*. SAGE, London.

Wynne, Brian (1996). Misunderstood misunderstanding: society identifies a public uptake of science. 19–46 in Alan Irwin and Brian Wynne (edited), *Misunderstanding Science? The public reconstruction of science and technology*. Cambridge University Press, Cambridge.

Wynne, Brian (2005). Risk as globalizing "democratic" discourse? Framing subjects and citizens. 66–82 in Melissa Leach, Ian Scoones and Brian Wynne (edited), *Science and Citizens: globalization and the challenge of engagement*. Zed Books, London.

Wynne, Brian and Sue Mayer (1993). How science fails the environment. *New Scientist*, 5th June, 33–35.

Zimmerman, Corinne and Kim Cuddington (2007). Ambiguous, circular and polysemous: students' definitions of the "balance of nature" metaphor. *Public Understanding of Science* 16, 393–406.

3

PARTICIPATING PUBLICS

Introduction

In 2010, I volunteered to be part of a Community Liaison Group based near my home in northern England. The group was set up by a private-sector company as part of preparing a planning application for a large municipal waste disposal and recycling facility nearby, after the company won a contract to manage the future municipal waste generated from two local authorities for the next twenty-five years. The group was described as an opportunity for local people to learn about the planning process, to have their views heard by the company and even influence the design of the eventual planning application, as well as to fulfil the company's aim of being a 'good neighbour' to local residents.

This Community Liaison Group met five times, called in various 'experts' to give evidence and to answer questions about the planning process and the environmental and health effects of the proposed facility. Initially, most of the local residents who attended the group meetings were very positive about being able to participate in planning something that would be a huge feature of their local area. But as time went on, they became increasingly disillusioned about whether the company was listening to them and whether what they said in group meetings would have any influence on the final development. After five meetings, several members of the group withdrew from the process, a planned sixth meeting was cancelled and a letter was written to the company and the local authorities from members of the group saying that they felt they had been misled into thinking that their participation would be taken seriously and have influence. Two years later, the planning application was approved and construction of the facility began in 2015.

Being involved in this long process really brought home to me the complexity and diverse expectations of environmental publics and their participating practices. The ten months of interaction raised contentious questions about who can participate, whom they can represent, how public participation in environmental decision-making is organised, experienced and resisted and whether participation has any power or influence.

This chapter looks at these and other diverse practices performed by publics participating in environmental decision-making and management, to explore how such practices are organised geographically and sociodemographically, how they are understood, carried out and resisted and what they can tell us about the power (and powerlessness) of environmental publics as they are imagined and produced.

Practices

I want to begin with the terms that are used, often interchangeably, for different practices that try to bring publics into discussion with environmental decision-makers. This is because the broad term 'public participation' can cover a variety of different practices and some similar practices are given different names.

A famous "ladder of citizen participation" was published in 1969 by Sherry Arnstein that is still relevant to this discussion today. It named eight different sorts of public participation practices and ranked them, with the best and most empowering practices at the top and the worst and least empowering at the bottom. The implication of Arnstein's ladder was that those organising public participation should try to climb it to the 'better', more democratically worthy practices on the higher rungs. As she put it:

> There is a critical difference between going through the empty ritual of participation and having the real power needed to affect the outcome of the process.
>
> *Arnstein 1969*

Although her ladder was not specific to environmental publics, it is still useful because it makes explicit that participation practices are very diverse and that they have implications for power (of which more later). Her ladder had eight rungs, thus:

8. Citizen control
7. Delegated power
6. Partnership
5. Placation
4. Consultation
3. Informing
2. Therapy
1. Manipulation

Arnstein considered that rungs 8, 7 and 6 on the ladder together provided "citizen power" where lay people, the non-elites or, in her words, the "have-not citizens", could make decisions, either alone or with others. Rungs 5, 4 and 3 were merely "tokenism" where the public may be listened to, but do not affect the decisions made by elites who retain power; even Placation, for example, may involve participation merely to "rubberstamp" decisions by elites. Rungs 2 and 1 she considered to be "non-participation", that is, merely the pretence of participation with the aim of "curing" publics of misinformation or ignorance about the "right" decision (as in the deficit model discussed in Chapter 2), but without listening to public opinion or being influenced by it. Rungs 2 and 1 also aim to "prove" that the public have participated by doing things (answering surveys, coming to meetings), in case this is disputed later or used against decisions.

Arnstein acknowledged that eight rungs simplified what might be hundreds of different practices. What is interesting is that, on the surface, practices on different rungs of the ladder may look the same, e.g. committee meetings may provide "Information" but also "Placation" by seeming to involve key members of local publics while the real decisions are made elsewhere, such as in the voting chamber of a local authority. It is not solely the format or design of the participation 'technology' that matters, but how it is used, that is, the practices that occur before, during and after it, such as chairing, handling questions and even organising furniture in the space to be used. Imagine a participation meeting in which the invited or accredited 'experts' sit on chairs on a raised platform at the front - this will influence how they are asked and answer questions from 'the floor', compared with a meeting where everyone sits in chairs arranged in a circle on the same level. Participation exercises thus sometimes make their imaginaries of publics and experts tangible through simple technologies of furniture and spatial arrangements, in an assemblage of meanings, objects and (assumed) skills (Truninger 2011, see Chapter 1). And participation practices go on not only during a participation exercise, but beforehand as it is designed and afterwards as it is evaluated and future consultations or actions are planned: exercises have life beyond the temporal restrictions that appear initially.

In the years since Arnstein's ladder was published, 'consulting' the public has become a common and necessary part of environmental planning and decision-making, but it remains arguable whether the practices have moved much up that ladder. There is frequently a commitment in policy to involving publics in decision-making, reinforced (in principle at least) by the Aarhus Convention in Europe, but this still may not be applied in practice (e.g. Irwin 2006; Petts 2005).

One reason is that the choice and purpose of participation practices are implicitly shaped by the imaginaries of the organisers, and such elites often continue to assume that 'the public' are ill-informed, biased or misinformed (e.g. Clifford 2013), reflecting the 'deficit model' in Chapter 2. But they may also assume that publics are driven by emotion rather than reason, highlighting that the deficit model also underpins a 'rationalist' ideal of public participation (Owens 2000, p. 1141), an ideal that aims to inform 'the public' and, by correcting any ignorance or misunderstandings, secure support for new technology or developments through encouraging rational, objective judgements, rather than what are assumed to be emotional, knee-jerk reactions.

As we saw in Chapter 2, the deficit model has been debunked by twenty years of social science research, showing that giving 'the public' information does not straightforwardly change people's views or behaviour (e.g. Horlick-Jones et al. 2007; Owens 2000). In this chapter, we need to take this a bit further to highlight that, as well as influencing knowing, ideally participation would "help not only to identify or implement solutions but to define, or reframe, what the problems actually are" (Owens 2000, p. 1144). In other words, true participation would not only appear near the end of a decision-making process to help with public roll-out, but would happen at the beginning of the process, to help design its agenda and the range of solutions it will consider.

Under this 'deliberative' or 'civic' ideal (Owens 2000), public participation is advocated both for reasons of democracy, that is, giving people the right to be involved in decisions that affect them or are paid for by their taxes, and for reasons of pragmatism, i.e. involving people should result in better, more acceptable decisions that can be more easily implemented. Another benefit is that, ideally, publics also learn through the process of participating (Petts 2001; Tippett et al. 2005), thus improving their own education, knowledge and social capital, which is also linked to environmental justice (Petts 2005; Walker 2011), and even having fun on the way as a positive effect.

With this ideal in mind, researchers have considered how best to design public participation exercises, evaluating them against criteria such as representativeness, legitimacy, effectiveness and impact (e.g. Maiello et al. 2013; Rowe and Frewer 2005; Tippett et al. 2005). Usually, weaker public participation is characterised as a one-way flow of information from the exercise designer to the public, whereas stronger public participation should facilitate information flows in both directions, going beyond consultation to achieve dialogue and deliberation "where participants actively develop and contribute to proposals, options, or evaluations during periods of extended involvement" (Burgess and Chilvers 2006, pp. 719–720). In a similar way, Dryzek et al. (2009) refer to participants in consensus conferences looking at GM foods as 'deliberative publics', reflecting movement up Arnstein's ladder towards publics sharing decision-making power with elites.

However, this remains an ideal and may neglect the negative effects of participation, such as apathy or protests against its validity (as in the case that opened this chapter). And there is another side to participation not captured by Arnstein's ladder: participation exercises – through their very design and technologies – themselves produce publics in the image constructed by their designers, that is, publics are also performed and produced through such exercises (e.g. Felt and Fochler 2010; Irwin 2001). Public participation exercises thus not only engage publics in more or less democratic ways, they also shape them as subjects of democracy, of information and of power. Warner (2002) suggested that particular or specialised publics are enacted through the way that they are 'addressed' by newsletters, webpages or other materials produced by those designing and conducting public engagement exercises. Hence, how a public is defined as the target or focus of a participation exercise is heavily influenced by what those designing that exercise want to achieve (Owens 2000, p. 1141; also Irwin 2001), which may not necessarily be democratic participation.

For example, local authority officers in the UK who consult local publics about air quality management have favoured 'one-way' techniques, such as information provision through leaflets, over more challenging 'two-way' techniques, such as the dialogue and Q&A format of more open public meetings. They have also tended to consult statutory stakeholders (e.g. government environment departments or health protection agencies) more than local publics, due to assuming that lay publics would not understand the complex issues involved (Dorfman et al. 2010), in line with the deficit model in Chapter 2. Such assumptions thus shape (and often limit) the type of consultation exercises and techniques that local authorities use and thus the type of publics that are involved and the influence they wield.

This interpretation reflects what Michael (2012, p. 531) called "the performative or ontological turn" recently in science and technology studies that sees the public "as emergent from the relations in which it is immersed and through which it is enacted". For him, these relations include identification (how "lay public citizenship or citizenliness" is established and validated), intervention (how the public voice is enabled through "formal mechanisms of voicing") and mediation (how the public voice is represented in scientific policy-making) (Michael 2012, p. 530). So, 'public imaginaries' shape how participation exercises are designed and constructed (Irwin 2006; Michael 2009; Warner 2002), but how these exercises are themselves enacted in turn shapes the publics with which they engage.

These 'public imaginaries' are also plural, because there are many notions of 'the public', some positive and some less so. For example, designers of public participation exercises often assume that 'the public' is excluded, powerless and environmentally under-informed (Maranta et al. 2003; Owens 2000; Ungar 2000), so when more knowledgeable publics do successfully engage in environmental debates, this often causes surprise and even hostility. As we saw in Chapter 2, the 2003 UK 'GM Nation?' participation exercises were criticised for being dominated by people who had pre-existing and highly negative opinions about genetically modified (GM) food and who were, as a result of expressing these opinions, not seen as representing 'the general public' because they were not neutral or open-minded on the issue.

The 'GM Nation?' debate was intended to allow the public to design the methods and scope of the debate about GM food, as an 'experiment' in extending democracy, but was instead seen by the policy community of elite policy-makers and scientists as failing to engage "with the broad mass of hitherto disengaged members of the lay public" (Horlick-Jones et al. 2007, p. 175) beyond those already involved in the debate and who had developed strong views or allegiances. This was because that policy community had constructed a notion of 'the public' as a neutral, open-minded 'general public', a *good public* that would listen to the arguments and come objectively to a balanced consensus, a notion that was itself flawed.

Good and bad environmental publics

This echoes the dichotomy identified in Chapter 2 between the ideals of 'good' and 'bad' publics that are imagined, enacted and sometimes resisted in participation exercises (Eden and Bear 2012; Gibson 2005; McClymont and O'Hare 2009; Walker et al. 2011). Table 3.1 summarises these ideals in the context specifically of participating practices and highlights several differences that I shall discuss in more detail.

First, Table 3.1 shows that publics are often expected to be representative of the wider population or 'general public', in terms of both their sociodemographic profile (age, gender, income, education level) and their opinions on the topic under debate. For example, in a public participation exercise for a council's waste management strategy analysed by Bull et al. (2008, p. 705), three 'community advisory fora' were set up, run by consultants but "independently chaired", with between sixteen and twenty people attending each who were "deemed to be representative of the general interests and experiences that existed in the community as a whole … not just made up of those already environmentally aware or those sympathetic to the proposals", a process that the authors described as a "highly innovative engagement process based on deliberative ideals" that was "largely effective". Here, very much echoing the 'GM Nation?' case, the ideal of a representative, unbiased public was used directly to design the exercise.

However, most participation practices are highly selective in recruiting participants. The people selected to become members of the Community Liaison Group mentioned in the opening of this chapter were not chosen on the basis

TABLE 3.1 Imagining good and bad publics in public participation

'Good' publics are imagined as…	*'Bad' publics are imagined as…*
Representative of the population or majority	Selective (and small) minority
Atomised individuals	Organised groups
Uncertain but neutral, so open to persuasion by 'rational' information	Not open to persuasion, because fixed in their views, often emotional in their reactions
Non-expert, not scientifically trained, largely ignorant of specialised issues before public engagement starts*	Knowledgeable before public engagement starts, perhaps contaminated by incorrect information or extreme opinions
Trusting of and willing to believe experts/elites	Sceptical and cynical of experts/elites
Cooperative with the participation process	Challenging or resistant to the participation process
Democratic because they represent the wider public interest	Undemocratic because they only represent their own self-interest or a shared but narrow group interest

Sources: Eden and Bear (2012), Irwin (2006), Irwin and Wynne (1995) and Michael (2009 p. 627).
Notes:
*Michael referred to them as 'pure', Irwin referred to them as 'innocent'.

of representativeness, disinterestedness or any other ideals from the participation literature (e.g. Rowe and Frewer 2000). Instead, invitations went out to local parish councils and other local groups and, via newsletters and websites, to local residents. Any such process results in self-selection, often by those with the most interest in and time available to spend on participation. But if a participation process is consequently seen as dominated by self-selected individuals biased by pre-existing opinions and experiences, it can be denigrated by those who do not like its outcomes – ideology intervenes, imaginaries shift and the publics that are involved are portrayed as *bad* publics.

In the example I used to open this chapter, representativeness was also referenced against 'the local community', another problematic term that conjures up an image of a cohesive or even homogenous set of local residents, glossing over divisions that may exist in age, race, income, interests or opinions. Staeheli et al (2009, p. 641) distinguished the term 'community' from the term 'public', arguing that members of a community share "a kind of social solidarity [that] derives from sharing a preexisting history, experience, or identity", whereas a public can "reach beyond those who are already known to each other. ... to draw strangers into discourse". Indeed, many public participation exercises seem to assume that participants will be strangers to each other, whereas many may be part of community groups and know (or know of) each other through local networks, as happened in the Community Liaison Group at the beginning of this chapter and in the 'GM Nation?' case. Moving participation exercises online is becoming increasingly common in recent years, which can increase the potential for atomised and even anonymised participation, although friendship networks and community groups still may be identifiable.

Second, organisers of public participation usually imagine the public to be neutral and therefore open to persuasion, showing how the 'deficit model' from Chapter 2 persists. Participation practices often focus on giving information to publics, answering questions from publics and countering 'misinformation' or 'misunderstanding'. Being neutral, being impartial and responding to information all also imply that the ideal environmental public should not be emotional, but instead 'rational'.

Third, and part of being imagined as open to persuasion and receptive to information, *good* publics are often also imagined to be *not* scientifically trained and thus 'lay', reflecting the persistent lay/expert dualism that we met in Chapter 2. Of course, the term 'expert' is very variable in terms of how it is defined in different circumstances – in a scientific laboratory, a court of law, a college classroom, for instance – and depending on what environmental issue is being discussed. But it may include doctors and other experts in public and environmental health, professional staff from environmental protection agencies such as river hydrologists, aquatic ecologists and atmospheric modellers. Formal qualifications, especially doctorates in scientific subjects, are often used to demarcate 'experts' from non-experts, that is, from 'the public', as discussed also in Chapter 2. For example, three Community Advisory Forums (CAFs) set up by Hampshire County Council for publics to participate in decisions about future waste management still used 'expert' witnesses to provide information (Petts 1997; 2001).

But expertise itself may be contested through participation practices. Opposing publics can seek out or construct and invoke their own 'experts', allowing them to counter-claim and set up expertise itself as an object of negotiation and contestation with the participation process. In the example that starts this chapter, some members of the Community Liaison Group identified and deployed their own 'expert' to counter the 'experts' put forward by the company and councils, such as planning officers, Environment Agency staff and Health Protection Agency staff. Instead, the members found a university professor in environmental chemistry through an online anti-incineration network and cited his work frequently, in paper reports, in online materials and in group meetings. In so doing, they did not set themselves up as experts, but they refused to accept that only the developer's 'side' had access to expertise – they also claimed this resource.

But this may go further if publics are themselves changed by participation in becoming more knowledgeable about the technical aspects of the debate, perhaps reading up them outside the 'official' participation events. Campaigning to defend one's local area "may be an instinctive reaction" (Rootes 2007, p. 732) but it also leads more consciously to learning and self-education about the issues. Individuals may have submitted their own planning applications for small building works to local authorities for approval, so have experience themselves of how planning process works – they are not therefore merely blanks to whom participation organisers offer information about larger, more contentious planning proposals.

Fourth, as we see in Table 3.1, publics can become less trusting of those identified as 'experts', often challenging the knowledgeability or competence of planning officers or other officials to make sensible judgements. Practices of participating are thus intertwined with active practices of knowing, as discussed in Chapter 2, emphasising again that knowing and doing are bundled together in practice. So, public opinion may not necessarily be an *input* to public participation exercises, but instead be an *output* (Irwin 2006), shaping new identities and constructions of the public. Unfortunately, the long-term and diffuse effects of this are difficult to trace and nearly impossible to prove, given the multiple other influences on people's opinions such as media stories (see Chapter 2) and how few studies continue to track participants and their views after participation exercises have closed.

Participation exercises can also be counterproductive for their organisers, prompting aggrieved or worried publics to set up new groups to challenge decisions controlled by elites. Consultants to waste companies that plan to build incinerators have therefore recommended that they set up "countermovement organizations … called Citizen Advisory Committees" (Walsh et al. 1993, p. 26), to work against the expected opposition from anti-incinerator movements. To put this another way, the companies are being advised to counter the effects of increasingly well-organised and well-informed ('bad') publics by setting up pro-incinerator ('good') publics that the companies can more easily control and use to publicise their arguments.

In some cases, however, it is not the 'general public' that is sought for public engagement exercises, but 'specialised publics', such as farmers being recruited to participate in agricultural policy or anglers and canoeists in river management. Specialised publics may also be (seen as) keener to become involved in decision-making than the general public, because their interests are more obviously affected and they are easier to identify and contact through named groups or clubs, although they may also be less likely to be persuaded of the state's viewpoint.

But specialised publics can also be seen as problematic because they are not representative of the general population, but are biased by their narrowly defined interests, whether these are sociodemographic, geographical or hobby-related. Recreational anglers may be considered to threaten democratic decision-making if their representatives are too vocal or override the interests of other river users in policy committees. The Environment Agency of England and Wales defines its flood management narrowly around the interests of publics who have been or are likely to be victims of flooding, leading the Agency more towards engineering solutions than other methods of floodplain management (Harries and Penning-Rowsell 2011).

Sometimes specialised publics are labelled as NIMBY, motivated by 'Not In My Back Yard' attitude to new developments and often very vocal in opposing such developments and dominating public participation (e.g. Clifford 2013). Labelling participants as NIMBYs is another example of constructing 'bad' publics with pre-existing opinions by contrasting them against the ideal of "the 'open-minded' (or 'innocent') citizen" (Irwin 2006, p. 315).

As a consequence of such imaginaries, Clifford (2013) found that professional planners imagined publics to be actively hostile, motivated by NIMBYism rather than the public good, being "defensive" and reluctant to attend participation events. Similarly, wind farm developers designed participation based on their previous experience with 'difficult' publics, assuming that "vociferous public opposition typically [was] in the minority whereas local supporters were the silent majority whose voice was not being heard" (Walker et al. 2010, p. 942). As well as constructing publics, this also constructs the identity, attributes and privileged position of the professions (the not-publics) that try to engage with them and constructs participation as a threat or danger to those professions, a danger that they seek to control through participation design (Clifford 2013; Walker et al. 2011). The environmental justice literature (e.g. Walker 2009, p. 626) identifies this as a part of a politics of stigmatisation and misrecognition, where participatory processes disrespect, denigrate and devalue some people (and their views) but not others, whether intentionally or through subconscious bias. Emphasising 'environmental justice' not only aims to show how environmental impacts of decisions vary by race, location and income, for example, but also to make the process of reaching those decisions fair by ensuring that participation is open to all, especially those who are likely to be impacted. But establishing what Walker (2009, p. 627) refers to as 'spaces of fair process' can be difficult in practice, especially where geographical proximity is used to define those who will potentially feel the impact of a decision because this proximity can be used to undermine the legitimacy of protests from those who live, work and play close to a controversial site by denigrating their concerns as NIMBY.

Sometimes, publics 'misbehave' in participation exercises (Michael 2012) in other ways, e.g. by failing to turn up, disrupting, talking about irrelevant issues, not taking exercises seriously or simply being present but silent. When they thus fail to fulfil the expectations of the 'good' public held by the exercise designers, they may be ignored or simply written out of the process.

So Table 3.1 summarises two opposing imaginaries of the public: one of an ill-informed but neutral public, open to persuasion through democratic processes; and one of a well-informed public open to involvement but also active on their own initiative, perhaps knowledgeable but already fixed in their views, sceptical about experts and potentially resistant to the participation process as undemocratic. The dualisms of good/bad participation and good/bad publics play out through these imaginaries and the way that public participation exercises attempt to control them, as we shall see later when we consider the implications for power.

Experimenting with practices

If the persistent ideal of participation is flawed, because it seeks to engage publics who are rational, disinterested, representative, little-informed but open to persuasion, what else can we try?

First, experiments to improve participation in environmental decision-making have sought to bring public participation 'upstream', that is, to involve publics much earlier in the process of developing new technologies or proposals, before decisions and substantial financial investments have been made.

Second, exercises have been designed to foster more two-way discussion between publics and elites. Examples are consensus conferences, where publics directly ask questions of 'experts' about contentious environmental or scientific issues and debate the answers, often writing a report to government on their conclusions. But even experimental forms of practice tend to import existing imaginaries of publics. Consensus conferences on plant biotechnology were run in New Zealand in 1996 and 1999, but both were biased because all the pro-biotechnology 'experts' giving evidence were scientists and all the anti-biotechnology 'experts' were not, so the lay panellists received the overall impression that there was a scientific consensus that was pro-biotechnology, an impression which proved difficult for them to challenge (see Goven 2003).

Third, participation exercises have also experimented with technology. For example, participatory computer mapping (P-GIS, see Cinderby 2010, Radil and Jiao 2016) aims to bypass the problems of traditional public engagement (including the exclusion or under-representation of 'hard to reach' groups in deprived communities) and to encourage public engagement through interactive mapping and visualisations, rather than wordy discussion. But even such well-intentioned experiments often struggle to redress inequalities and differences in the publics involved, in terms of income, race, education and so on, which result in some groups being under- or over-represented in the participation process and thus their views being under- or over-represented in the maps that result (e.g. Radil and Jiao 2016). Online participation exercises also have the potential to innovate in recruiting, engaging and interacting with more people, although they are not without problems and have so far been little researched.

Fourth, and more unusually, experiments sometimes directly reflect or are inspired by cutting-edge research. After a destructive earthquake near Constitución in Chile in 2010, groups were set up to involve the local publics in planning and re-building the city and these groups were explicitly designed to reflect academic research in science and technology studies, specifically a book by Callon, Lascoumes and Barthe (2009) that the participation coordinator had recently read and that advocated 'hybrid forums' for open discussion by a heterogeneous mix of experts and publics (Tironi 2015, p. 576). Despite this seemingly impeccable pedigree, the local publics in Constitución failed to fulfil the coordinator's expectations, partly due to deep class divisions, partly due to their immediate needs for food, clothing and supplies far outweighing their interests in planning the longer-term future of their city and fulfilling their civic duty through participating in that process. Hence, few 'ordinary' residents attended and the groups were instead populated mainly by representatives from NGOs or civic, industry and union leaders.

So the sad fact is that most public participation exercises have still not moved up Arnstein's ladder much in terms of actual practice over the last few decades, despite upholding the ideal of open, two-way, influential participation in principle (Irwin 2006; Serrao-Neumann et al. 2015). The imaginary of a neutral, 'public', capable of being informed and persuaded to change their minds, remains persistent, but illusory.

Place

I now want to look more explicitly at the geographies of public participation, which means examining the roles of place, space and scale in shaping environmental publics through participation.

In situ *and* ex situ *practices*

First, participation practices are often *ex situ* practices, that is, they take place away from the site of the decision, e.g. people meet in town halls to discuss ways to manage river flooding, focus groups are convened in local pubs or hotels to discuss planning applications for incinerators, as in the case that opened this chapter. The place in which participation happens is thus more likely to be controlled by the elites designing the participation exercises and may well be unfamiliar to those visiting it for the first time for the purpose of participation. This may generate emotional reactions which have little to do with the issue at stake and more to do with the intimidating effects of architecture and the hospitality of strangers, especially when representatives of large corporations provide free food and slick presentations.

Participation spaces also have their own microgeographies. As mentioned above, the spatial arrangement of chairs and raised platforms at a venue shape how participation can (and cannot) happen there. Again, this reflects the concerns of the environmental justice literature with fairness and effectiveness beyond the spatialities involved in recruiting participants, especially from locally impacted populations, to the spatialities involved in the ongoing practices of those participants.

The outside environment is then brought into the setting through technological props. Environmental data are selected, displayed and authorised by reference to professional landscape assessors or to scientific models, instrumentation and researchers, rather than by reference to local people and their habits. Photomontages are constructed to visualise future landscapes and the impacts of new developments in a seemingly straightforward way that is immediately comprehensible to any (sighted) layperson without the need to interpret numbers or text, while also rendering invisible the in/exclusions performed during their construction as part of standardised techniques for visual impact assessment.

In contrast, sometimes participating publics are taken on field trips to the sites where the environmental decision will be implemented, for *in situ* discussion. For example, the Environment Agency has invited local publics affected by (and indeed very angry about) flooding in the English Somerset Levels in 2013–14 to visit pumping stations and see how the river water is managed at those sites. Members of the Community Liaison Group mentioned at the beginning of this chapter were invited to visit (at the company's expense) existing waste disposal facilities nearby, to see for themselves how they worked. Such visits have the potential to open up the geographies of participation, to enhance transparency and improve learning through embodied and observational processes, as well as discursive ones. However, the elites managing the site being visited usually retain control over what visiting publics can and cannot see and will inevitably be selective in how they present themselves and their site management.

When the issue is city planning, it is often easier to hold participatory meetings *in situ*. For example, participatory GIS meetings aiming to map out citizen concerns in urban areas are often held in those urban areas. However, even here it is possible that the geographical selection of the meeting places may reflect the organisers' knowledge of the city or town, especially regarding where suitable meeting places are and whether people can travel to these places safely and easily.

As discussed in Chapter 2 in relation to the location of learning, the location of participation is therefore diverse and shaped by the technologies of its surroundings. But the immediate visibility of environments to publics participating *in situ* may be more influential than the tabling of photomontages of a new factory in a room *ex situ*, far away from the proposed development site. Geographies of participation design thus continually (re)shape their own outcomes.

Scaling practices and interests

As well as the site or place of participating publics, the scale of participating practices is highly significant, even more so than in the previous chapter. There are several aspects to this. First, there is the scale at which the organisers of participation exercises operate, whether they be national government (as in 'GM Nation?'), local government (as in the planning authority in the case that opens this chapter), multinational companies that operate local sites (as in the company in the same case) or local people organising their own events in their neighbourhoods. All of these influence the ways in which participating practices are designed, implemented and given meaning.

Second, there is the scale of the decision in which publics are invited to participate. The 'GM Nation?' exercise declared its national scale in its very title, as it was supposed to feed into the national government's decision on whether to allow genetically modified crops to be grown in the UK. Other decisions may be made locally and result in new buildings or other developments locally, but also contribute to environmental change globally, e.g. a new incinerator that was approved by a local authority will also emit carbon dioxide into the global atmosphere. And 'the public good' is often invoked as a scale-less but clearly wide-ranging possible justification for a variety of environmental decisions, from preventing building on floodplains to expanding airports.

But scaling practices are also highly contentious. Geographers have shown that scales are not pre-given nor do they simply reflect objective measures of space (such as km^2). Rather, scales are constructed, ordered within hierarchies (especially local, regional, national, international, global) and deployed to demarcate power and influence in decision-making. In imagined geographies of participation, scales are often defined in oppositional terms, so whether a practice is linked to claims of (or is accorded to) a global scale or a local scale can shape perceptions of it and limit its influence. Often, the national scale is invoked as having greater political and ideological authority than the local scale, which is associated more with personal experience and accorded lesser influence.

Using scale to organise geographical thinking becomes very problematic when scale is used as shorthand for the spatialization of other political, social and economic ideas (Collinge 2006; Marston et al. 2005). This "discursive power" of scale (Marston et al. 2005, p. 420) is used to shape the subjectivities of people and the specific characters of places. For example, the global scale of the economy or of climate change may be seen as more important than local scale changes in employment levels or climate risks. As a corollary, the higher-scale changes may be assumed to be *driving* the lower-scale changes, such as global changes in climate driving local shifts in weather or flooding, national changes in policy or new forms of waste management being driven by the economy, rather than the different scales being dynamically interlinked. Using scales somewhat differently, Walker et al. (2011) distinguish between publics 'in places' and promoters of renewable energy technologies 'in networks', suggesting that publics are more readily imagined as geographically embedded in the local context than companies and other promoters of renewable energy, as the latter are more 'footloose'.

So, constructs of scale that are deployed in participation practices shape those practices by downplaying some arguments and giving more credence to others. Normative loadings of scale, their inherently hierarchical ordering and its consequences for assumptions about power, influence and control disturbed Marston et al. (2005) so much that they argued that we should dispense with the idea of scale entirely and instead think in terms of the extent (areal size) and the borders of different spaces in a 'flat ontology'. Critics of this view argued instead that the idea of scale cannot be simply dispensed with; rather, we should interrogate scales more explicitly as multiple and dynamically constituted through socio-political struggles (e.g. Jonas 2006) and rethink them as emergent and performative (e.g. Collinge 2006).

For the purposes of this book, I see scales not as given and immutable, but as imagined and changeable attributes of practices. Scale is performed through spatializing practices (e.g. Collinge 2006). For example, when inviting publics to participate in a planning exercise, spatialization is implicit in choices about where to post invitations, which media outlets (local newspaper, parish magazine, town website) to advertise on and how to frame geographically the invitation in words and images. As well as shaping environmental publics, invoking scale produces governance objects related to specific scales, e.g. river basins, community forests (Cohen and McCarthy 2015). And although the scale of public participation exercises may often be defined by traditional political spaces, e.g. a nation-state's decision over allowing GM crops, a local authority's decisions over allowing a new incinerator or housing development, the scale of environmental change rarely maps readily onto these political spaces, whether they are ecological zones affected by genetically modified species or global climate changes affected by increased carbon dioxide emissions from burning waste.

To challenge assumptions about hierarchical scaling, Devine-Wright et al. (2015, p. 76) used the concept of 'place attachment' to develop "a relational approach to multiple forms of belonging". In their survey, respondents who said they believed that climate change was human-induced were more likely to be 'attached' to (that is, feel they 'belonged' to) the global scale of the planet than to the local scale of their neighbourhood or city. In contrast, people who said that they felt they 'belonged' more to their country than to their neighbourhood or the planet were more likely to say the human contribution to climate change was lower and less certain, and more likely to oppose climate action. Devine-Wright et al. (2015) therefore argued that people can be attached to global-scale places such as the planet in the form of climate change, as well as to local ones, challenging the 'localist' tendency to assume that people value things and landscapes that are nearer to them rather than ones that are further away.

Imagining local publics as NIMBYs

Like the idea of 'the public', then, the idea of 'scale' can also be constructed or imagined in both positive and negative ways. A good illustration of this is the way that scale is used negatively to belittle the protests of environmental publics about local decisions or proposals. Here, the scalar hierarchy mentioned above, with global at the top and local at the bottom, is deployed to denigrate protests about new developments that are seen as locally anchored, such as proposed wind farms, homeless shelters and incinerators.

As I mentioned above, such protests are often labelled as 'NIMBY' to imply that those who protest about very local changes are not driven by ideals of altruism, civic interest in the public good or even wider ideologies such as environmentalism or social justice. Instead, they are assumed to be solely driven by their self-interest, and their 'Not In My Back Yard' attitude is motivated by the desire to defend their own environment against loss of amenity, falls in house prices or unwanted 'others' that such new facilities might bring, with little regard for whether successful protest merely pushes a development elsewhere and its impact onto another neighbourhood. Understanding local concerns as merely the "selfish parochialism of local residents" (Gibson 2005, p. 384) evokes geographical determinism by portraying such residents as *bad* environmental publics (see Table 3.1), using an explicitly scalar shorthand to try to undermine their claims.

Such negative labelling can be very powerful when deployed by professional elites. As we saw above, planners can use this as a defensive strategy, portraying the planning professional as championing the views of the "silent majority", the polar opposite of the minority of vociferous NIMBYs (Clifford 2013, p. 126). Like the deficit model we met in Chapter 2, the notion of NIMBYism has been debunked by scholars over recent years because it is simplistic (McClymont and O'Hare 2008), because it fails to explain the true roots of opposition, being unproven empirically (Wolsink 2006) and because it is asymmetrical in the way that local *opponents* of a new facility are accused of bias due to (localised) self-interest but local *supporters* of the same facility are not. The label NIMBY is therefore a geographical term of abuse to belittle lay publics living close to new developments who oppose those developments, although there is no comparable term to belittle lay publics living close to new developments who support those developments. Indeed, sometimes those living closest to a facility may be more positive about it than those living further away (Devine-Wright 2009, e.g. Sellafield nuclear power station, Macgill 1987; Wynne et al. 1993/2007), meaning that invoking geographical determinism or local protectionism can be inaccurate as well as limiting. NIMBYism also often implies that local publics are ignorant or misinformed about the impacts of a proposed facility, whereas they are often highly informed from both local experience and also because they pay more attention to media coverage of their local area (Devine-Wright 2009), so that "protesters are not NIMBYs at all, but rather have good and often wide-ranging grounds for their complaints" (Burningham 2000, p. 58).

It has been therefore argued that researchers should abandon the term NIMBY entirely (Burningham 2000; Wolsink 2006) or that "so-called 'NIMBY' responses should be re-conceived as place-protective actions, which are founded upon place attachment and identity" (Devine-Wright 2009, p. 432). Later, Devine-Wright (2013) argued that place-attachments may not be solely local but 'polyscalar', that is, an individual might have diverse attachments to places at different scales, whether a local park or a nation-state. However, this approach does not reject the scalar hierarchy as such, but retains the notion of a global-scale attachment to 'one world' as the best way to support climate change mitigation. To put it another way, if a change to the local environment can prompt strong reactions and protests, how can we find a way to use changes to the global environment to prompt such reactions and protests as well?

The corollary to such powerful imaginaries is that, often, environmental publics become aware of the threat of being labelled NIMBYs, and deliberately invent, name and portray themselves in ways to avoid this. What Michael (2009) called 'Publics-in-Particular' thus reflexively produce themselves, both as distinctive from 'the general public' and as having particular knowledges, identities and interests (also Lezaun and Soneryd 2007). For example, Walsh et al. (1993, p. 36) contrasted two areas where protestors were arguing against proposals for a new incinerator, and suggested that those who were more successful in "ridding themselves of the 'NIMBY' tag by emphasizing the importance of serious recycling and the proposed incinerator's negative consequences for those living further from the site" were also more successful in having the incinerator rejected from their area. In the case of the incinerator with which I began this chapter, participants were very conscious of the risks of being labelling as NIMBYs and worked very hard to 'upscale' their opinions to the wider region when speaking in participation practices (see Chapter 6 for more details).

The other side of the NIMBY coin is that specialised publics can also use their local knowledge to construct themselves as sources of information, not merely recipients of it. Place location can therefore be portrayed (and imagined) more positively: "living locally is at the same time an important resource for objectors, providing the basis for claims of local knowledge and experience which are often used to counter the assessments of 'experts' who do not live in the area" (Burningham 2000, p. 61). This is one reason to develop and apply participatory GIS to enhance participation through mapping and visualisation, rather than discourse, but can still result in such exercises being seen as locally scaled, rather than of wider interest (e.g. Radil and Jiao 2016). Scale is not only imagined in the process of constructing and framing public participation exercises, it is also imagined in the process of joining, conducting and resisting such exercises, helping to (re)shape those environmental realities.

This is even more pronounced when participation moves online, using internet surveys, comments boxes, emails and social media. Online practices for protesting and campaigning have already attracted a lot of media and academic attention, but online practices for participating in planning and other environmental decisions has been less researched so far. This may be because state-led planning exercises lack the speed and responsiveness that characterise social media, and thus have become rather entrenched in more traditional, offline ways of working. Whatever the reason, shifting to online participation may stretch geographies of participation further, even re-scaling some aspects of the process such as recruitment, although many others aspects may still be geographically delimited.

Power

All of the participating practices that I have discussed so far have implications for how power is associated with and enacted. Indeed, Arnstein (1969, p. 216) argued that the two are inseparable:

> citizen participation is a categorical term for citizen power. It is the redistribution of power that enables the have-not citizens, presently excluded from the political and economic processes, to be deliberately included in the future. It is the strategy by which the have-nots join in determining how information is shared, goals and policies are set, tax resources are allocated, programs are operated, and benefits like contracts and patronage are parceled out. In short, it is the means by which they can induce significant social reform which enables them to share in the benefits of the affluent society.

What strategies and practices are used to organise, allocate, commandeer and direct power to or away from participating publics? As we saw above and in Chapter 2, despite arguments for broader and more innovative modes of participation, attempts at involving the wider public (especially attempts sponsored by the state) still frequently assume that 'the public' is excluded, powerless and unknowledgeable about the environmental issue in question (Maranta et al. 2003; Owens 2000; Ungar 2000). Improved public engagement is therefore advocated to correct these problems and move towards more open, inclusive environmental governance for various policy issues (e.g. Horlick-Jones et al. 2007; Irwin 2006; Irwin and Wynne 1996; Maranta et al. 2003; Owens 2000; Petts and Brooks 2006; Ungar 2000; Walker et al. 2010).

But the existing imaginaries of the public continue to be very powerful in shaping how the state or other agents approach public engagement (e.g. Ellis and Waterton 2004; Michael 2009; Warner 2002) and the resulting engagement exercises can themselves contribute to defining and organising 'the public' in limited ways (Felt and Fochler 2010). For example, the offer of 'community benefit' funding by wind farm developers (e.g. Walker et al. 2010) constructs local people not only as recipients of information but also as recipients of monetary compensation.

Control over how engagement is framed is thus "an important source of power" (Irwin 2006, pp. 315–6). Inclusion and exclusion are performed through invitation practices where organisers define which publics are invited to participate and how, and through self-selection practices in terms of who chooses to turn up to a participation exercise. These are also highly spatialized practices, as we saw above. The environmental justice literature (e.g. Walker 2009) thus identifies the un/fairness of participation procedures as an expression of procedural justice during decision-making, which is distinct from the distributional in/justice of the eventual environmental impact of the decision made.

And participation may be further framed in terms of how exercises are designed and performed by the publics who do choose to attend. For example, deliberative debates by invited publics are usually chaired by someone and it is the chair's job to keep the debate civil, focussed and democratic (as far as possible, in terms of allowing different people to be heard), as well as to finish debates on time. Although often implicit, the choice of the chair and the way that chairing is performed will inevitably shape the outcomes of the debate, how participants feel about their participation and whether this influences them to participate again, thus changing the character of those publics. Again, this reflects the procedural aspects of environmental in/justice during decision-making, which in turn influence the distributional in/justice of the final decision through its environmental impact on different publics. And those publics who do turn up and experience this chairing may go along with it or resist it by being rowdy, not reading the materials requested, going 'off-message' and seeking to suppress or outdo those with viewpoints that they disagree with. There are all practices that respond to and enact power relations.

Another common framing practice is to design participation too narrowly around discursive and representational practices of talking, writing and mapping. These might include some discussion of how people feel, but sometimes those feelings cannot be easily articulated in a discursive manner that is acceptable within the participation format. What publics know (see Chapter 2) and how they well they can articulate their knowledge and views in words, rather than where they live or what they do, becomes a key factor in shaping which publics participate in environmental decision-making and how.

Public imaginaries are constructed through opposition and dualism and actively shape engagement. Defining the public therefore matters for environmental democracy and public involvement (Felt and Fochler 2010; Staeheli and Mitchell 2007; Walker et al. 2010), because it influences how (and with whom) participation designers and state agencies seek to engage. The continued use of information as the main way to involve people, the rejection of public views where they are not neutral and how elites attempt to monopolise science and fact – these qualities of the deficit model persist and emphasise that public participation often remains an exercise in talking the talk but not walking the walk. And even well-intentioned public participation exercises can fail to achieve consensus when the views of commercial developers are irreconcilable with the views of local communities because they emanate from very different values and ideologies, differences that cannot be removed through providing information or debate (Rootes 2007, p. 733).

So far I have concentrated on discursive or rationalist modes of engagement, that is, public participation that takes place in offices and other indoor spaces of debate and constructs environmental publics primarily through written (Warner 2002) or spoken debate (Felt and Fochler 2010; Michael 2009; Lezaun and Soneryd 2007), whether online or in person. Such discursive modes of participation often prioritise cognitive knowledge, lay expertise and interaction with the state and its agencies (e.g. Barnes et al. 2003).

A very different way to imagine participation by environmental publics is through directly managing environments, that is, rather than debating how environments should be protected or changed through a meeting in a town hall, some publics get out into those environments themselves and manage them through embodied practices of planting, digging, dredging and so on. Thinking of environmental publics as participating by doing, rather than by thinking and arguing, is rare in the literature but exemplifies the arguments in this book. For example, angling clubs in England manage the river reaches where they retain fishing rights through volunteer labour and practical actions (see Chapter 5; Eden and Bear 2012). As 'hands-on' lay managers, these recreational anglers plant or remove vegetation, add or remove species, repair or destroy structures such as weirs and groynes, erode or accumulate sediments in waterways. On land, the Woodland Trust in England recruits local volunteers to plant trees to expand woodland through embodied labour. Environmental realities are thus (re)made by environmental publics as they directly participate not merely in the talk of environmental management, but also in its practice.

Summing up

Public participation exercises are often designed around an imagined public, that is, a public imagined by the designers to be neutral, poorly informed but open to new information and persuadable. But where the participants fail to fulfil this ideal of a good public in practice, they can be cast as badly behaved publics, as no longer legitimate representatives of the 'general public' and as undermining even well-organised and well-populated public participation.

So traditional approaches remain problematic, especially where they rely on the 'deficit model' of public understanding that we met in Chapter 2 (see Irwin 2006; Irwin and Wynne 1996; Michael 2002; Wynne 1995). Public engagement can be ineffective or counter-productive if debates are framed in such unhelpful ways (Irwin 2001) or if engagement is geared to seeking legitimation for state policy and not to generating genuine and significant public input to decisions.

Felt and Fochler (2010) argued that, if we think of public engagement exercises as technologies for levering stakeholder input, then we should consider not only how these exercises are designed but how also they are used. Such an approach is helpful because use is not merely passive – users dynamically shape technologies even as usage also performs its users, meaning that public definitions and views are fluid, as we have seen in this chapter. Amorphous and ambiguous multiple constructions of the public are therefore held by different groups, including members of the public, as "laypeople enact themselves as particular sorts of publics" (Michael 2009, p. 620).

Scaling is also practised to include and exclude particular publics and to define which publics are seen as legitimate by deploying scalar hierarchies of worth, especially local-national-global and the negative labelling of NIMBY. And the dichotomy between 'good' and 'bad' public participation, between disinterested general publics and self-interested local publics, compliant publics and resistant publics, may be invoked explicitly or implicitly to shape how participation exercises are designed and experienced.

Public engagement exercises are often geared to rather grand policy decisions or future events, not the more mundane and everyday practices of environmental management at a local scale. But while people may know local, familiar environments very well and value them accordingly, they may find it difficult to relate abstract environmental policy concepts (like sustainability or climate change) to their everyday realities (Harrison and Burgess 1994; Macnaghten 2003; Myers and Macnaghten 1998). And holding participation events *ex situ* in anonymous office blocks or hotel meeting rooms, a common practice, does little to acknowledge the geographical contexts being debated nor the ways in which local publics feel and know about local environments under discussion. This 'de-placing' of public participation is another spatial strategy for control, a geographical fix that seeks to neutralise emotional engagement, but one that is rather neglected in the literature.

And even where publics participate fully, they may have no effect on environmental decisions. Reflecting Arnstein's Tokenism or Placation (see above), public participation in technological and environmental decision-making may still serve primarily to legitimate decisions, rather than to build dialogue between the state and its publics, often creating disillusionment amongst those publics who did participate and potentially discouraging them from doing so again (e.g. Irwin 1995).

So practices of engaging environmental publics in debate and decision-making encompass both knowing and doing, being and behaving, imagining and acting, discursive and embodied practices. The design of engagement exercises is shaped by (perhaps unacknowledged) expectations of the public and how they should behave, expectations which are frequently confounded by the kinds of publics who participate in practice and the practices that they perform. This participation also changes publics by altering how they are performed and imagine themselves. And participation also goes beyond discursive practices to hands-on environmental management as a way for publics to participate directly in decision-making and also to carry those decisions out. Meanings, practices and spaces are all shaped through these encounters to enact a far more messy and problematic set of relationships between people, place and power than we see in the democratic ideal, but relationships that perform everyday social life, as well as present and future environments.

References

Arnstein, Sherry (1969). A ladder of citizen participation. *AIP Journal* 35, 4, 217–224.

Barnes, Marian, Janet Newman, Andrew Knops and Helen Sullivan (2003). Constituting "the public" in public participation. *Public Administration* 81, 2, 379–399.

Bull, Richard, Judith Petts and James Evans (2008). Social learning from public engagement: dreaming the impossible? *Journal of Environmental Planning and Management* 51, 5, 701–716.

Burgess, Jacquelin and Jason Chilvers (2006). Upping the ante: a conceptual framework for designing and evaluating participatory technology assessments. *Science and Public Policy* 33, 10, 713–728.

Burningham, Kate (2000). Using the language of NIMBY: a topic for research, not an activity for researchers. *Local Environment* 5, 1, 55–67.

Callon, Michael, Pierre Lascoumes and Yannick Barthe (2009). *Acting in an uncertain world. An essay on technical democracy.* The MIT Press, Cambridge, MA.

Cinderby, Steve (2010). How to reach the "hard-to-reach": the development of Participatory Geographic Information Systems (P-GIS) for inclusive urban design in UK cities. *Area* 42, 2, 239–251.

Clifford, Ben P. (2013). Rendering reform: local authority planners and perceptions of public participation in Great Britain. *Local Environment* 18, 1, 110–131.

Cohen, Alice and James McCarthy (2015). Reviewing rescaling: strengthening the case for environmental considerations. *Progress in Human Geography* 39, 1, 3–25.

Collinge, Chris (2006). Flat ontology and the deconstruction of scale: a response to Marston, Jones and Woodward. *Transactions of the Institute of British Geographers* 31, 244–251.

Devine-Wright, Patrick (2009). Rethinking NIMBYism: the role of place attachment and place identity in explaining place-protective action. *Journal of Community & Applied Social Psychology* 19, 426–441.

Devine-Wright, Patrick (2013). Think global, act local? The relevance of place attachments and place identities in a climate changed world. *Global Environmental Change* 23, 61–69.

Devine-Wright, Patrick, Jennifer Price and Zoe Leviston (2015). My country or my planet? Exploring the influence of multiple place attachments and ideological beliefs upon climate change attitudes and opinions. *Global Environmental Change* 30, 68–79.

Dorfman, Paul, Dave C. Gibbs, Nurul Leksmono, James Longhurst and Emma Louisa Caroline Weitkamp (2010). Exploring the context of consultation: the case of local air quality management. *Local Environment* 15, 1, 15–26.

Dryzek, John S., Robert E. Goodin, Aviezer Tucker and Bernard Reber (2009). Promethean elites encounter precautionary publics: the case of GM foods. *Science, Technology & Human Values* 34, 3, 263–288.

Eden, Sally and Christopher Bear (2012). The good, the bad, and the hands-on: constructs of public participation, anglers, and lay management of water environments. *Environment and Planning A* 44, 1200–1240.

Ellis, Rebecca and Claire Waterton (2004). Environmental citizenship in the making: the participation of volunteer naturalists in UK biological recording and biodiversity policy. *Science and Public Policy* 31, 2, 95–105.

Felt, Ulrike and Maximilian Fochler (2010). Machineries for making publics: inscribing and de-scribing publics in public engagement. *Minerva* 48, 219–238.

Gibson, Timothy A. (2005). NIMBY and the Civic Good. *City & Community* 4, 4 December.

Goven, Joanna (2003). Deploying the consensus conference in New Zealand democracy and de-problemization. *Public Understanding of Science* 12, 4, 423–440.

Harries, Tim and Edmund Penning-Rowsell (2011). Victim pressure, institutional inertia and climate change adaptation: the case of flood risk. *Global Environmental Change* 21, 188–197.

Harrison, C. and Burgess, J. (1994). Social constructions of nature: a case study of conflicts over the development of Rainham Marshes. *Transactions of the Institute of British Geographers* 19, 291–310.

Horlick-Jones, Tom, John Walls, Gene Rowe, Nick Pidgeon, Wouter Poortinga, Graham Murdock and Tim O'Riordan (2007). *The GM Debate: risk, politics and public engagement.* Routledge, London.

Irwin, Alan (1995). *Citizen Science.* Routledge, London.

Irwin, Alan (2001). Constructing the scientific citizen: science and democracy in the biosciences. *Public Understanding of Science* 10, 1–18.

Irwin, Alan (2006). The politics of talk: coming to terms with the "new" scientific governance. *Social Studies of Science* 36, 2, 299–320.

Irwin, Alan and Brian Wynne (1996). Conclusions. 213–221 in Alan Irwin and Brian Wynne *(edited), Misunderstanding Science? The public reconstruction of science and technology.* Cambridge University Press, Cambridge.

Jonas, Andrew E.G. (2006). Pro scale: further reflections on the "scale debate" in human geography. *Transactions of the Institute of British Geographers* 31, 3, 399–406.

Lezaun, Javier and Linda Soneryd (2007). Consulting citizens: technologies of elicitation and the mobility of publics. *Public Understanding of Science* 16, 3, 279–297.

Maiello, Antonella, Ana Carolina Christovão, Ana Lucia Nogueira de Paiva Britto and Marco Frey (2013). Public participation for urban sustainability: investigating relations among citizens, the environment and institutions – an ethnographic study. *Local Environment* 18, 2, 167–183.

McClymont, Katie and Paul O'Hare (2008). "We're not NIMBYs!" Contrasting local protest groups with idealised conceptions of sustainable communities. *Local Environment* 13, 4, 321–335.

Macgill, Sally (1987). *The Politics of Anxiety.* Pion, London.

Maranta, Alessandro, Michael Guggenheim, Priska Gisler and Christian Pohl (2003). The reality of experts and the imagined lay person. *Acta Sociologica* 46, 2, 150–265.

Marston, Sallie A., John Paul Jones III and Keith Woodward (2005). Human geography without scale. *Transactions of the Institute of British Geographers* 30, 416–432.

Michael, Mike (2009). Publics performing publics: of PiGs, PiPs and politics. *Public Understanding of Science* 18, 5, 617–631.

Michael, Mike (2012). "What Are We Busy Doing?" Engaging the idiot. *Science, Technology, & Human Values* 37, 528–554.

Myers, G. and P. Macnaghten (1998). Rhetorics of environmental sustainability: commonplaces and places. *Environment and Planning A* 30, 2, 333–353.

Owens S. (2000). Engaging the public: information and deliberation in environmental policy. *Environment and Planning A* 32, 1141–1148.

Petts, Judith (1997). The public–expert interface in local waste management decisions: experts, credibility and process. *Public Understanding of Science* 6, 359–381.

Petts, Judith (2001). Evaluating the effectiveness of deliberative processes: waste management case-studies. *Journal of Environmental Planning and Management* 44, 2, 207–226.

Petts, Judith (2005). Enhancing environmental equity through decision-making: learning from waste management. *Local Environment* 10, 4, 397–409.

Petts, Judith and Catherine Brooks (2006). Expert conceptualisations of the role of lay knowledge in environmental decision-making. *Environment and Planning A* 38, 6, 1045–1059.

Radil, Steven M. and Junfeng Jiao (2016). Public participatory GIS and the geography of inclusion. *The Professional Geographer* 68, 2, 202–211.

Rootes, Christopher (2007). Acting locally: the character, contexts and significance of local environmental mobilisations. *Environmental Politics* 16, 5, 722–741.

Rowe, Gene and Lynn J. Frewer (2000). Public participation methods: a framework for evaluation. *Science, Technology, & Human Values* 25, 1, 3–29.

Rowe, Gene and Lynn J. Frewer (2005). A typology of public engagement mechanisms. *Science, Technology, & Human Values* 30, 2, 251–290.

Serrao-Neumann, Silvia, Ben Harman, Anne Leitch and Darryl Low Choy (2015). Public engagement and climate adaptation: insights from three local governments in Australia. *Journal of Environmental Planning and Management* 58, 7, 1196–1216.

Staeheli, L. A. and D. Mitchell (2007). Locating the public in research and practice. *Progress in Human Geography* 31, 792–811.

Staeheli, L.A., D. Mitchell and C.R. Nagel (2009). Making publics: immigrants, regimes of publicity and entry to "the public". *Environment and Planning D: Society and Space* 27, 633–648.

Tippett, J., B. Searle, C. Pahl-Wostl and Y. Rees (2005). Social learning in public participation in river basin management: early findings from HarmoniCOP European case studies. *Environmental Science & Policy* 8, 287–299.

Tironi, Manuel (2015). Disastrous publics: counter-enactments in participatory experiments. *Science, Technology, & Human Values* 40, 4, 564–587.

Truninger, Monica (2011). Cooking with Bimby in a moment of recruitment: exploring conventions and practice perspectives. *Journal of Consumer Culture* 11, 37–59.

Ungar, Sheldon (2000). Knowledge, ignorance and the popular culture: climate change versus the ozone hole. *Public Understanding of Science* 9, 297–312.

Walker, Gordon (2009). Beyond distribution and proximity: exploring the multiple spatialities of environmental justice. *Antipode* 41, 4, 614–636.

Walker, Gordon (2011). *Environmental Justice: concepts, evidence and politics*. Routledge, London.

Walker, Gordon, Noel Cass, Kate Burningham and Julie Barnett (2010). Renewable energy and sociotechnical change: imagined subjectivities of "the public" and their implications. *Environment & Planning A* 42, 931–947.

Walker, Gordon, Patrick Devine-Wright, Julie Barnett, Kate Burningham, Noel Cass, Hannah Devine-Wright, Gerda Speller, John Barton, Bob Evans, Yuko Heath, David Infield, Judith Parks and Kate Theobald (2011). Symmetries, expectations, dynamics and contexts: a framework for understanding public engagement with renewable energy projects. 1–14 in Patrick Devine Wright (edited), *Renewable Energy and the Public: from NIMBY to participation*. Earthscan, London.

Walsh, Edward, Rex Warland and D. Clayton Smith (1993). Backyards, NIMBYs, and incinerator sitings: implications for social movement theory. *Social Problems* 40, 1, 25–38.

Warner, M. (2002). Publics and counterpublics. *Public Culture* 14, 49–90.

Wolsink, M. (2006). Invalid theory impedes our understanding: a critique on the persistence of the language of NIMBY. *Transactions of the Institute of British Geographers* 85, 85–91.

Wynne, Brian, Claire Waterton and Robin Grove-White (originally 1993, updated 2007). Public perceptions and the nuclear industry in West Cumbria. www.csec.lancs.ac.uk/docs/Public%20Perceptions%20Nuclear%20Industry.pdf

4

CONSUMING PUBLICS

Introduction

In 2007, the designer Anya Hindmarch designed a shopping bag made of unbleached cotton that bore the slogan "I'm not a plastic bag". Sold for £5 in Sainsbury's supermarkets with a limit of thiry bags per store, supplies of the Hindmarch bag reportedly sold out within an hour of going on sale, prompting stories that 'green' had become fashionable, that "sustainable is the new black" (BBC 2007).

This was a high-profile UK example of the burgeoning global movement to stop retailers giving out and consumers accepting millions of free plastic single-use carrier bags in shops, to reduce unnecessary use of environmental resources and to avoid the bags ending up as litter that is both unsightly and potentially damaging to wildlife. The Hindmarch bag was a product designed to replace multiple other products and thus reduce resource use, but its own environmental footprint was considerable: it was made from cotton and shipped across the world from China. And the bag itself promoted more consuming through its own function – not only did consumers buy the Hindmarch bag, but they also bought things to put into it. Despite replacing something given away for free, the Hindmarch bag became a desirable consumer product, contrasting with the usual arguments that green products are disadvantaged by these types of 'green premium' on prices, e.g. on organic food or free-range eggs.

Elsewhere, legislation has been adopted to change consuming practices. Bangladesh banned plastic single-use carrier bags entirely because they were clogging up drains and exacerbating flooding (Clapp and Swanston 2009) and Wales began to charge 5p per bag in 2011, with use falling 71 per cent by 2014 (Welsh Government 2014). But because many people used the free, supposedly 'single-use' bags not only for carrying their shopping home, but also for lining their rubbish bins or wrapping a picnic lunch, some have argued that legislation to reduce the production and consumption of some bags will also increase the production and consumption of others, e.g. as consumers start buying 'long-life' bags like Hindmarch's or other plastic bags specifically produced and marketed for lining rubbish bins.

The practices of environmental consumption are therefore not only diverse but sometimes contradictory, even downright puzzling for the analyst. This chapter is about these practices of consuming and how they relate to the making of environmental publics. Much of 'social practice theory' that I introduced in Chapter 1 (e.g. Shove 2010; Warde 2005, 2014) developed from the study of consumption in sociological terms, partly because of the challenge of analysing such multiple, repeated, mundane but still puzzling routines of everyday life. This underpins much of what I discuss in this chapter, but my focus is more specialised than all of everyday life: it is about those consuming practices that relate to the environment.

Practices

One might argue that all consumption is inherently environmental, in the sense that buying and using things, whether a shopping bag, a laptop computer, a washing machine or a cooking stove, uses resources in some way, whether material resources like fabric or oil or more intangible ones like electricity.

And this is further complicated by the threefold meaning of 'consumption': buying, using and using up (i.e. exhausting, depleting) are all possible interpretations of consuming practices. They all matter in environmental terms, but may differ across space, time and participants: one person may buy food in a shop but then take it home where another person may cook the food, perhaps a week or a month later, for somebody else to eat and yet another different person may recycle the tins or compost the vegetable peelings some days later. At different times, people may cook the same food in different ways; they may read the labelling on the package even as they chew through the contents at the dinner table (thus consuming both the information and the nutrition at the same time, combining the material with the intangible) or they may evaluate its quality or branding and decide not to purchase it again. There is not a singular moment of consumption, but a process, an encounter that may be short or long and cover different spaces (shop, home, picnic site) as the consumer and the food product travel through this heterogeneous network. Things are commodified and "decommodified ... through a series of moments, of which exchange for money is just one" (Sayer 2003, p. 346).

In this chapter, I focus more on pro-environmental consumption practices and publics, that is, consumption practices that are deliberately and consciously thought about as environmental and how this makes environmental publics in different places. In cradle-to-grave analyses that attempt to measure the overall environmental impact of a product as it is produced, consumed and disposed of, we often find that consuming is greater not in practices of purchasing but in practices of using. For example, by far the greatest environmental impact of a domestic washing machine is caused during use, rather than manufacture or even disposal (WRAP 2010), due to the amounts of water and electricity used to wash clothes, week in and week out.

So, this chapter is not solely about buying things, but also about consuming things after purchase or even *before* purchase, e.g. in the case of electricity that is billed after use. With such a wealth of diverse practices to cover, there is a lot to pack into this chapter, but I will emphasise how environmental publics are made through these practices even as they enact them.

Purchasing practices

Since the 1980s, market researchers have been trying to measure how many 'green' consumers there are and to classify them by sociodemographics, geodemographics and other distinguishing characteristics. This means that, unlike some of the other practices in this book, practices of consuming have been well defined and intensively studied, because they have become part of the commercial business of retailing and market research.

There have been many attempts to measure the 'green market' of consumers who buy environmentally themed products, e.g. the Cooperative Bank's 2012 survey, see Table 4.1.

There are several problems with such measurements. First, they may seem to be about purchasing practices, but they are not – they are about reporting and self-representing practices because they rely on people saying that they will or do buy 'green' products, that is, they measure 'self-reported' behaviour or expressions of intent, rather than actual purchasing. Surveys also ask about intention – not what publics do, but what they might potentially do. In a TNS survey for the EU (Eurobarometer 365 2011, p. 74), 72 per cent of respondents across Europe (ranging from 59 per cent in Portugal to 89 per cent in Sweden) said that they were "ready to buy environmentally friendly products even if they cost a little bit more" and 25 per cent said they were not. This conjures a large, potentially profitable 'green market' into being, that is, an imagined public that companies could be encouraged to sell to and expand further through appropriate advertising

TABLE 4.1 People reporting various ethical consuming behaviours at least once during 2011

% reporting	Ethical choices and attitudes
50%	Avoided product/service on basis of company's responsible reputation
42 %	Bought primarily for ethical reasons
41 %	Recommended a company on basis of company's responsible reputation
33 %	Actively sought information on company's reputation
31 %	Felt guilty about unethical purchase
24 %	Actively campaigned on environmental/social issues

Source: The Cooperative Bank (2012).

and product innovation. Hence, such measurements are not solely motivated by environmental reasons, but often by economic ones as well.

Second, measuring consumption tends to focus on consuming at the shop counter or sales website, but the spatialisation of consuming is usually far broader and more complex than this, including where and how we browse for, purchase, use and eat products, as I shall show later. Methods of measurement often fail to encompass these multiple places and modes of consuming.

Third, measuring consuming through surveys or observational practice is often done on a 'one-off' basis that does not consider the regularity of such practices or the kinds of products, spaces and relationships on which they depend. So buying an environmentally themed product once a month would be counted the same as buying them every day in the TNS example above, yet would have vastly different implications for how consuming is given meaning and how much environmental impact is caused. We need to acknowledge more explicitly that consuming practices are temporalized differently by place and type: in the home, electricity, water and other services are used continually, food stocks are bought sporadically, sometimes on impulse, but consumed (in the sense of being eaten) every day and expensive items like computers and fridges are bought less frequently but perhaps with far more thought in advance.

Environmental consumption may also rise and fall over time in tandem with wider economic trends, becoming relatively well established, that is, normalised in some sectors but remaining prone to change in others. For example, in the UK, ethical products initially bucked the recent global recession as sales grew 9 per cent from 2009 to 2011 (The Cooperative 2012) and by 2014, the whole ethical market was estimated to be worth £38 billion; with ethical investment added, this estimate rose to £80 billion (Ethical Consumer Research Association 2015). But some types of ethical products seemed to do better than others: in the UK, sales of sustainably sourced fish rose to £462 million in 2013–14 (partly due to UK celebrity chef, Hugh Fearnley-Whittingstall), but other ethical sectors diminished, with sales of organic food barely keeping ahead of inflation and Fairtrade products seeing their first fall in sales for twenty years (Ethical Consumer 2015).

The distinction between environmental and ethical reasons is worth considering briefly. There are different terms used, including 'ethical consumption' (e.g. Barnett et al. 2010), 'political consumption' or 'political consumerism' (e.g. Micheletti and Stolle 2008) and 'sustainable consumption' (e.g. Soper 2007), covering many possible motivations or consumer rationales. Although much of the sociological and geographical literature tends to focus on boycotting products associated with exploitative labour practices or oppressive regimes, especially the high-profile case of FairTrade, ethical consumption in its widest sense can also include veganism, eating only kosher or halal meat, favouring products from a particular country (usually one's own), boycotting products from a particular country (usually *not* one's own) and buying investment portfolios that screen out links to organisations supporting abortion, contraception or other contentious practices. 'Ethical' thus relates not only to environmental or left-wing concerns, but can include any aspects of production and profit that consumers consider important.

One difference between ethical consumption and environmental or sustainable consumption is that the latter considers not only the type of products consumed but also the total amount of consumption: environmentally speaking, what matters is not merely what is consumed and how, but how much. But measuring 'how much' we consume is tricky and again varies by sector. For instance, in the UK it seems that environmental publics are increasingly consuming more food but less energy in recent years. The Department of Health (2015) estimated that people are getting heavier, with 62 per cent of adults and 28 per cent of children aged 12–15 years old classified as overweight or obese, while the Department for Energy and Climate Change (2015) estimated that people are using less domestic energy, the amount falling 19 per cent from 2000–14 (37 per cent since 1970), helped by warmer winters in 2011–14 and despite the population rising by 10 per cent in the same period.

Fifth, consuming also involves choices not to consume, so we also need to consider deliberately *non*-consuming practices, such as boycotting. Again, attempts to measure this have been made, e.g. in a UK survey for Defra (Thornton 2009, p. 42):

- 46 per cent of people surveyed said they would be prepared to pay more for environmentally-friendly products and 31 per cent said they would not.
- 30 per cent said they were not buying some products because of too much packaging.
- 21 per cent said they were choosing to buy wood products certified as from sustainable sources.
- 19 per cent that they were buying peat free compost.
- 73 per cent said they made an effort to buy things from local retailers and suppliers and 11 per cent said they did not.

These results mix together choices to buy and choosing not to buy, as well as environmental, ethical and economic ('buy local') reasons. The Ethical Consumer Research Association (2015) estimated that 20 per cent of the UK population boycott specific products or outlets for ethical reasons, with lost sales to those companies of £2,640 million in 2014.

Using practices

As well as buying and not buying, the consuming practices of environmental publics also involve using practices, both before, during and after purchase. Electricity, gas, oil, wood and water are used in homes across the world, becoming routinized as part of the infrastructure of domestic life (e.g. Shove 2003), rather than being special purchases that are consciously deliberated. Other products are routinely used in times and spaces distant from the act of purchase, e.g. petrol for cars, clothing and holidays. Practices of environmental consuming are therefore about more than purchasing; they are also about using.

This emphasis on using has prompted some explicitly practice-based analyses of domestic consumption of resources, such as water for showering (Shove 2003) and electricity for running appliances (Gram-Hanssen 2011), which consider the implications of such using practices for sustainability and environmental impact (also Hargreaves et al. 2013; Spaargaren 2011). For example, when analysing smart metering of electricity use in fifteen households, Hargreaves et al. (2010) found that a key motivation for participating in this study was that householders were interested in learning more about their energy use, linking practices of consuming and knowing. Also, although the households varied as to how much their energy use declined after the meters were installed, one consequence was that some new habits arose around and through meters (e.g. turning off appliances), and became so routinized that they no longer depended on the meters themselves: householders stopped looking at the meters but continued the changed practices that had been prompted by the meters when newly installed.

This reminds us that environmental publics are shaped or 'co-produced' in part by devices and information, and in part also by their very participation in practices that then shift from conscious and deliberate choices to unconscious, perhaps unarticulated habits. For example, the manual for my washing machine says that it is highly efficient (A++ rating) but only if the drum is full of dirty clothes; half-loads are not efficient. Because I read this, I now often add items that are not very dirty to a wash, in order to fill up the drum and make the washing more efficient, but this may use more water and energy overall. As well as new technological practices of consuming (e.g. turning off or replacing appliances), Hargreaves et al. (2010) found some households also reporting new or changed social practices of discussing energy use with family members, again emphasising how consuming practices change individuals and their relationships as they are enacted.

Similarly, how we use products such as a car, usually an infrequent and expensive purchase in comparison to water or energy, is illuminating. Ozaki et al. (2013, p. 522) applied practice theory to how a hybrid (that is, a potentially 'sustainable') car is actually used by drivers, to analyse how devices actively take part in practices and coevolve with their users. They analysed the Toyota Prius, a car with a built-in computer that tells the driver how many miles are driven per gallon of fuel and thus enables (and even encourages) the driver to change how they are driving to improve fuel efficiency. Drivers could thus gain pleasure and satisfaction from consuming *less* through driving more frugally, but some drivers in their study still sought to do the opposite and consumed more fuel. Resistance and 'subversion' can therefore persist even in the face of a seemingly sustainably motivated purchase, and one that is so infrequent and expensive that the decision is usually deliberated over far more explicitly than other everyday purchases. Using something therefore involves different practices from buying something:

> Sales of a supposedly sustainable technology do not automatically produce
> sustainable effects. Users might be motivated to purchase their products by

cost-based incentives, but that does not mean that such a purchase automatically engenders anticipated environmental effects. There is no linear relationship between environmental demand and supply ... effects results from evolving, and embodied interactions between the producer, technology, and user

Osaki et al. 2013, p. 537

Again we can link *using* practices with *not-using* practices, because sustainable consumption is often advocated in order to reduce overall energy use, especially through buying and using more efficient appliances. Defra (Thornton 2009) reported that 76 per cent of respondents in a 2009 survey said that they were cutting down on the use of gas and electricity at home (compared to 58 per cent in 2007) and 46 per cent said that they 'never' leave the lights on after they leave a room (11 per cent said that they 'always' or 'very often' did). This may explain why, as mentioned above, energy consumption in UK households has fallen in recent years, despite the number of appliances in households increasing. So the consuming practices of *buying* appliances may be inversely related to the consuming practices of *using* electricity to power those appliances, thereby negating any environmental benefits.

A further problem is that energy publics are not directly choosing their energy practices; rather, their energy use is embedded into (and thus rendered invisible by) appliance design or operation. In other words, people do not consume energy as such – they consume the services that it enables, e.g. entertainment, social networking or cooking dinner, with energy consumption indirectly supporting those social practices of consumption. For example, consuming energy by leaving electrical appliances on 'standby' mode influences not only the amount of energy consumed in homes, but also how environmental publics shape practices and are shaped by those practices. Many appliances, especially televisions and computers, now automatically go to 'standby mode' when they are not used for a set period of time – this is sometimes also called 'low power mode', 'sleep' or 'hibernation', and means that the appliance appears to be switched 'off' but continues to draw electricity, albeit at a lower level than when fully 'on' and being actively watched or used.

Standby mode is convenient for consumers because it keeps the clocks inside appliances running and makes devices quick to start up again, but it also wastes energy and increases the environmental impact of consumption. Up to 9–16 per cent of all domestic electricity consumption is now devoted to running devices on standby, used up in being idle. This is why standby power is sometimes referred to as 'vampire' or 'zombie' power, because it sucks electricity, usually overnight, for no purpose and people largely remain unaware of it. Because most electricity is still produced using fossil fuels (35 per cent from coal in the UK in 2012, according to the Department of Energy and Climate Change 2012), this vampire power is also carbon-loaded, thus contributing to climate change in future.

The vampire power consumed during standby also sheds light on how environmental publics are made through practice. First, more products are reverting automatically to standby mode after a period of inactivity (when consumers' attention moves elsewhere), so that standby mode is becoming the new 'norm' for products. In some cases, standby itself becomes a selling point in supporting quick re-start and new models are being produced with much higher average standby consumption, some over 17 watts, precisely to make them quicker to start up, an attribute that is often used in advertising.

But this is not only about products: if the next generation of publics are always online or wired up, if for them standby is the new norm and the off-switch is archaic or unknown, then their choice to be/use/consume otherwise is removed and this loss is first naturalised, then forgotten. In the Defra survey (Thornton 2009) quoted earlier, 52 per cent of respondents said that they 'never' leave a TV or PC on standby for long periods of time at home,[1] but 20 per cent said that they 'always' or 'very often' did. Since then, the latter is becoming the norm: as standby is embedded into our ways of living, so do practices and publics co-produce each other, with people feeling 'always on' (Turkle 2011) even as their devices are, forgetting alternative ways of working offline or even how to 'switch off' electrically and psychologically. Consuming is shaped not only by product qualities, but also by understandings of what constitutes 'off' and routinisation of standby makes electrical appliances phantoms by default, drawing down 'vampire' energy through normalised but hidden or forgotten practices that also shape how environmental publics use them.

As a consequence, energy consumption is often invisible, meaning that consumers often neglect or misunderstand the impact of their own practices. Gram-Hanssen (2011, p. 70) gives the example of a family in Denmark who liked fresh air and would open doors and windows a lot; they therefore assumed that their energy use would be lower than their neighbours and were surprised to be told it was higher than average amongst their neighbours. Analysing practices of using as well as consuming adds to but also complicates our understanding of how environmental publics and their practices develop.

As well as embedded systems like electricity in our homes, there are other consuming practices that are shaped by infrastructure and the policies of large organisations that serve us (e.g. Seyfang 2005). Public-sector institutions like schools and hospitals procure food and devices and even organise transport systems, thus shaping our individual consuming practices in addition to our own domestic arrangements. Consuming services through collective provision can be restrictive for individual consumers, e.g. fast food restaurants' inability to provide a burger without onions for onion-allergic Leigh Star (1990) to eat, because of the rigid systems of assembly. And in the case of electricity, although many suppliers offer 'green tariffs', consuming these differentiates practices economically but not materially, because electricity from different generation methods (gas, coal, oil, renewable, nuclear) is pooled through a national grid to domestic devices of all sorts. Again, choices and practices are rendered invisible by these collective provisions.

Wasting, divesting and recycling practices

As well as buying and using, consuming can also be thought of as wasting, that is, how a consumer decides to dispose of the remains of a product, or even an intact but unused product that is no longer wanted, is itself part of consuming practices.

Consumer waste is often regarded as highly correlated with development, i.e. wealthier societies usually buy and thus have more stuff at home, but they also may waste more stuff. This especially applies to food: an estimated 15 million tonnes of food are thrown away in the UK every year (more than the amount of packaging), 8.3 million tonnes of it by domestic households (22 per cent of the amount purchased), including 1.4 million bananas, 1.5 million tomatoes and 24 million slices of bread, but about half of the food thrown away, which amounts to 4.2 million tonnes worth £12.5 billion, could have been avoided because it was still edible (WRAP 2015, 2012). Despite these staggering statistics, Defra (2009) reported that 49 per cent of people surveyed said they threw away 'no' or 'very minimal' amounts of uneaten food, 45 per cent said they threw away a 'small amount' or 'some' and 5 per cent 'quite a lot' or 'a reasonable amount.' Clearly, self-reported practices in surveys fail to accurately reflect actual practices in homes, schools and workplaces.

And there are multiple ways to get rid of stuff from our homes: 'divestment' practices (Gregson et al. 2007) include leaving items outside our homes to be collected by local authorities, taking items to recycling points like bottle banks, 'handing down' items to friends and family, donating items to charity shops or school jumble sales and composting raw food peelings and other leftovers in our gardens. These practices also involve retailers, local authorities, charities and extended social networks, through complex geographies of movement and recycling, geographies that we shall examine in more detail later in this chapter.

Consumer recycling has risen in the UK since 2007, due to the landfill tax and local authority targets, bringing the UK's previously dismal recycling levels nearer to (if still lower than) those elsewhere in Europe. Recycling practices have become normalised in the UK, pushed into the mainstream through kerbside collections being promoted by local authorities and visibly enacted by neighbours as they put recycling boxes on pavements weekly or fortnightly. Other disposing and divesting practices, from charitable donation to composting, remain voluntary. To put some figures on this, in Defra's survey of people in the UK (Thornton 2009, p. 47):

- 91 per cent said that they recycled items rather than throwing them away.
- 84 per cent said that they took a shopping bag when shopping (rather than accept a free single-use bag from a shop).
- 76 per cent said that they reused items like empty bottles, tubs, jars, envelopes or paper.
- 43 per cent said that they composted household food and/or garden waste.
- 29 per cent said that they checked if an item could be recycled before they bought it.

In their survey, Defra were clearly interested in the differences between practices, but they missed an important one. Their survey suggests that three times as many people say they recycle as say they check if a product is recyclable before buying it, but Defra did not ask how many people *bought* recycled products, thus 'closing the loop' of their own practices. This implies not only that some recycling and reusing practices are normalised more than others, but also that people are perhaps more conscious of environmental aspects when they are disposing of a product than when they are purchasing it, a temporal disjuncture that illustrates the difficulties of assuming that practices transfer easily between contexts.

In some cases, divestment practices can be directly prompted by the desire or expectation to buy new things, thus facilitating future consuming by making more space in the home. This again links consuming as buying with consuming as disposing, through deliberate input-output calculations by householders that will vary depending on people's own circumstances, such as life-stage, housing tenure and changing employment or income.

To illustrate this, consider Freecycle,[2] a version of the more familiar eBay where people offer items that they no longer need or want, e.g. baby clothes, furniture, televisions, paint, garden plants and books, through online groups and someone who wants them can arrange to collect them. Freecycle groups aim to facilitate recycling and reuse, e.g. one group that I joined said on its website:

> Freecycle is an international grassroots environmental movement designed to minimise the amount of "stuff" we dump in our landfills. We do this by providing a forum for people to give their unused items to new, loving homes.

But while recycling, Freecyclers also imagine practices of 'good' consumption through avoiding landfill, repairing broken items, re-using and restoring old items and turning unwanted items 'surplus to requirements' in one household into wanted, usable and re-valued items in another household. Freecycling thus constructs "an *experiential space* that is actively produced and negotiated through both the discursive practices of its creators and the cognition of its surfers" (Currah 2003, p. 9, italics in original). The virtual space of Freecycle is also explicitly regulated as a non-commercial space, where no money changes hands, a form of 'alternativeness' to the mainstream that echoes my earlier discussion of ethical consumption. This is enforced through the (n)etiquette of Freecycling publics that seeks to exclude commercial dealers in second-hand goods who try to get stuff for free from Freecycle and then sell it on for profit; moderators repeatedly post reminders like this:

> Please don't advertise anything which you expect to get money for, and please don't use this list to advertise events ... NO POLITICS, NO SPAM, NO MONEY.

Consuming publics are thus shaped by their Freecycling practices (both online and offline), by taking on new items, by clearing out their own house, by considering different practices of use and repair, but also by being part of a group that continually discusses what is valuable and what is not, what is acceptable and what is not, what is normal and what is positively disruptive of the norm. Consumers thus reinforce, challenge and shift their disposing practices continually, through both materially exchanging products and imaginatively repositioning and repurposing products through online wordings, linking together consuming, using and disposing.

Prosuming practices

We could also see Freecycling as a form of production by consumers, what is often called 'prosumption'. Originating with Alvin Toffler in 1980 (Ritzer and Jurgenson 2010; Zwick et al. 2008), the term 'prosumption' emphasises how consumption increasingly blurs into production in retail environments. In self-service restaurants, supermarkets and petrol/gas stations today, consumers do (for free) the work previously done by paid employees, such as clearing tables, selecting food and fuelling cars (Ritzer and Jurgenson 2010). Online prosuming practices involve consumers making videos for YouTube or writing reviews of music, film and books on retailer websites like Amazon, effectively helping the retailer to sell products. Videos, blogs and advice sites originally produced for fun (as a form of leisure consumption) have become sufficiently popular to generate substantial revenues from advertising and some popular blogs have been turned into material products such as printed books, which people can then buy in 'bricks-and-mortar' stores. Through such practices, production by consumers helps to generate profits for commercial companies, although commentators disagree about whether prosumption empowers consumers or merely serves capitalist accumulation (cf. Currah 2003; Ritzer and Jurgenson 2010; Zwick et al. 2008).

Prosumption fits well into my arguments in this book, because it shows how environmental publics are made through consuming and producing practices. We might extend the idea of prosumption beyond these big commercial concerns to other examples, such as consumers making and selling their own products through eBay or their own (small) website. Defra (Thornton 2009) reported that 33 per cent of respondents who had a garden said that they grow their own fruit and/or vegetables (although probably in small quantities) and people who grow food on their allotments or in community gardens to eat at home often give any excess to their neighbours or sell it to raise funds for themselves or local charities.

Sometimes, consumers do not do the work to produce food themselves, but share their land with others who provide labour. In the USA, people can hire somebody to turn their backyard/garden into a productive vegetable plot, doing some (or no) work of pruning, planting, harvesting themselves, according to inclination, but still benefiting from eating any produce, as well as sharing it with the labourer. For some commentators, this is problematic because it effectively contracts out ethical/local consumption; for others, it enables self-production through advice and start-up help, fostering and empowering those practising prosumption, being seen by some garden owners as part of 'alternative' food movements that seek to avoid or even challenge the industrialised food supply system (e.g. Naylor 2012).

These examples emphasise that consuming practices are not separate from, but heavily influence producing practices, and that labour and love come together in many forms of leisure and household consumption. Having said that, how environmental are some of these practices? There is little investigation into whether Freecycling households and people with allotments live more sustainable lives and have lower environmental impacts overall than those who do not participate in these practices. So these publics and their practices may be couched in environmental terms and give meaning to their consuming by reference to ethics or sustainability, but they may not necessarily be reducing their overall environmental impact.

Learning to consume

Implicit in much of this chapter so far is that consumption practices also link to knowing practices. Consumers learn about products, services and ways to access both through various channels, some of which are themselves consumed, e.g. when someone buys a book like *The Green Consumer Guide* that lists products that are less environmentally damaging and therefore worthy of buying, or surfs a travel website about a country prior to visiting it on holiday, encountering various pop-up adverts.

These examples convey information separately from the physical products with the intent to shift consuming practices towards more sustainable ones, but there is also a great deal of on–product information available to consumers in the form of labels. Products may carry a logo or information to demonstrate, for example, that they have been certified by the Soil Association as organically produced, by the Forest Stewardship Council as produced from sustainably managed forests, by the Marine Stewardship Council as harvested from sustainably managed fisheries or by the Rainforest Alliance as produced through sustainable agriculture. Such labels serve as proxies for information that the consumer cannot access themselves, because they cannot visit forests and fisheries to check that they are well managed, not environmentally damaging and not over-exploitative.

Such information is also increasingly being provided online through ethical and sustainable consumption apps that can scan barcodes on product packaging while the consumer shops in a 'bricks-and-mortar' store and then rate it against specific criteria of being, for example, sustainably produced, fairly traded or low carbon. These digital proxies perform similar functions to the paper information on labels, although they may provide more detail and more immediate and interactive consuming of information.

And as well as labels on and information about consumer products, live metering technologies are also being rolled out to help consumers learn about their electricity usage, with the aim of shifting their practices towards sustainability and carbon reduction. To make more visible the hidden forms of consumption, such as water and electricity use at home, rather than receiving a bill every few months, consumers can watch the watts or litres tick away on a display unit, enabling them to experiment with the effects on the display of changing their embodied household practices, such as showering or switching lights on and off, and thus the effects on their environmental impact.

But we learnt in Chapter 2 that consumers do not necessarily react to information, even to reduce their bills, never mind to reduce their environmental impact. Hargreaves et al. (2010) interviewed fifteen households who had tried new electricity meters and many still said that they felt that they had little 'control' over what appliances they used and how, whether because of family habits, illness or their appliances being designed without an easy 'off' switch, as I discussed above. Consumers also found it difficult to make sense of the abstract, intangible units used by their meter, such as kilowatt hours or tonnes of carbon dioxide emissions. Also, although most households used the meters a lot initially, checking them regularly to see how much electricity they were using and through which appliance, this practice soon faded for some, either because the meter was not fitted in an easily viewable place, so the extra effort to go and look at it worked against routinisation, or because they internalised much of the information and, having learned which appliances were using more energy by studying the meter, they routinely switched those appliances off without needing to check the meter regularly.

Learning practices thus often support consuming practices, prompting changes but also raising questions for consumers, rather than answering them. This again reflects the deficit model from Chapter 2 in that learning does not necessarily change consuming, but the practices of getting, interpreting and judging information can shape consuming publics and their practices in unexpected or unstable ways, such as buying new devices to track electricity use, or using computers (and electricity to power them) to search online for energy-efficient products. So learning is both consuming and also shapes consuming.

Some would argue that, especially where encouraged by governments and large corporations, these various smart meters and proxy certification schemes merely encourage consumers to take responsibility for reducing their own environmental impact, reducing pressure on the state to take this responsibility through legislating, for example, so that cars are more fuel efficient. In this sense, environmental publics may be invoked as carriers and changers of practices through everyday decisions, rather than the big decisions made by the state.

And of course, sometimes the information that is consumed and acted upon is simply wrong. In October 2015, a scandal broke across the world when regulators accused Volkswagen of installing software in its diesel cars that allowed them to defeat official emissions tests, hiding the fact that the cars produced up to 40 times the US legal limit of pollutants (BBC 2015). 11 million cars bought by consumers under the Volkswagen, Audi, SEAT and Skoda brands were implicated (Volkswagen Group UK 2016) and Volkswagen expects to pay up to €6.5 billion to recall and fix the fraudulent cars, a bill that may well rise if 'class action' lawsuits by consumers are successful. Consumer groups such as Which? in the UK had for some years said that car manufacturers have falsely claimed better environmental performance than their vehicles achieved in practice, meaning that even if consumers receive, understand and act on environmental information about products, their practices may be more environmentally damaging than they expect.

I bought a second-hand SEAT car in February 2015, largely because of the very good fuel consumption that it promised. It turned out that my car is one of the 11 million with misleading software that needs to be fixed, so even by doing my 'research' into the various brands when deciding which to buy, I was misdirected. This certainly made me question my own ability as a consumer to enact my own choices, faced with a barrage of technical information and relying only on the manufacturer to ensure this is accurate and actionable. Not only can this make environmental publics far more sceptical about practices that are promoted as ethical or sustainable, such as buying fuel-efficient cars, but also undermines their own perceptions of themselves as knowing.

Place

So far, I have examined various consumption practices, only a few of which are actually about purchasing: more are about using and disposing of things and services. I will now move on to consider more explicitly the geographies of these consumption practices and how environmental publics are made by placing and scaling practices.

First, consumption is often associated with the private sphere of social practice, hidden away in a host of small, everyday decisions, prompted by individual desire and self-interest, distinguished from the public sphere of citizenship that is more positively associated with democratic practices such as voting (see Chapter 7) and participating in planning and environmental decision-making (see Chapter 3). For example, Hoppner and Whitmarsh (2011, p. 59) classify most consumption practices that impact on climate change, like using domestic lights, walking/cycling/driving/taking public transport, recycling, choosing food and buying products, to be in the 'private-sphere' engaging solely and individualistically with 'the market'. They then classify voting, campaigning or protesting, joining an NGO and participating in policy consultation as part of the 'public-sphere', because these engage with the state and other citizens directly.

This approach produces a strong and asymmetrical binary of value between public and private space, a binary which is often mapped onto another – and far more antagonistic - binary of citizen versus consumer practice (Soper and Trentman 2008, p. 2), as the next section also shows. Both binaries can then be used to theorise power relations and especially to denigrate consuming practices as less morally worthy, less effective and more delusional than other participatory practices such as protesting or voting (see Chapters 6 and 7 respectively).

Second, geographies of consuming practices map out across the world in ways that are complex but often obscure. The globalisation of production has encouraged footloose industries to move factories to wherever they find cheap labour, resources or minimal regulation and the internet has allowed commodities to be ordered online from any location and then shipped across the world to the consumer's home. It could be argued that such perspectives discount the spatialisation of consumption, rendering it aspatial or at least no longer bounded by political borders or the physical geography of transit routes across oceans and continents, being able to bring the 'remote' and the 'exotic' to everyone's doorstep. Online practices increasingly depend upon geo-locating and geo-referencing all sorts of information, from pinning Facebook posts on a Google map to apps using the location of a smartphone to list restaurants, tourist attractions or weather in the vicinity of the phone's user. This has been referred to as "the death of cyberspace through the revenge of geography" (Rogers 2012, p. 194), meaning that, rather than making cyberspace 'placeless' and ordered through topology and people's shared interests rather than geography, online networks are increasingly located in geographical space, e.g. the town at the centre of a Freecycle group. Geographies of consuming persist through online practices, rather than being wiped away.

And those geographies of consuming involve vast movements across space as millions of commodities are moved, exported, imported, repackaged and distributed across the landscape, with environmental impacts in terms of the air pollution and carbon emissions. Commodities do not only travel *to* consumers in shops, in courier vans and in the post; they also travel *from* them through disposing practices, transporting waste across the globe to places where cheaper labour costs turn dismantling and reuse into profit e.g. for ships (Crang et al. 2012).

All these hidden geographies of consuming are important but neglected. Geographers in particular have drawn attention to the 'distance problem' of consumption, that is, consumers have been physically separated from production by modernisation, industrialisation, globalisation and automation, disconnecting them from the environmental, social and political impacts of production, because the impacts happen far away but also because they happen behind the closed doors of the factory or shop. This renders the products 'innocent' to the consumer's gaze, which is a problem because they cannot tell us their stories: "the grapes that sit upon the supermarket shelves are mute; we cannot see the fingerprints of exploitation upon them or tell immediately what part of the world they are from" (Harvey 1990, p. 422).

Food is a very good example of these arguments, because fewer people are directly employed in agriculture than ever before and because increasingly urbanised societies, industrialised agriculture, complex global supply chains and new technological processes (e.g. genetic modification) also mean that fewer people know how food is produced or how to grow their own food, despite food being an essential component of everyday life. 'Distance' and 'disconnection' are thus both cartographic and cognitive: (rural) food producing spaces are assumed to be geographically remote from (urban) processing, retailing and consuming spaces but, even where people live next door to farms or food factories or walk past them every day, they still have little to do with and know very little (or nothing) about what happens inside.

So, modern geographies of food are argued to distance consumers from producers and hide the very social relations and environmental impacts that make food production possible (Duffy et al. 2005; Princen 1997; Hudson and Hudson 2003). Applying a political economy perspective to this, some have argued that consumers today are alienated from the conditions of production, particularly from the exploitation of labour remote from consumption, an exploitation that is then obscured by advertising and marketing, making products seem 'innocent' by stripping them of signs of their origins in pollution and hard labour (e.g. gold, see Hartwick 1998).

It is clear that such arguments conceptualise consuming publics as *unknowing* publics, at least in environmental terms, especially around food production processes. This parallels the deficit model from Chapter 2, which has been used to criticise how scientific communication invokes simplistic models of 'the public' (Gregory and Miller 1998), often imagining them to be helplessly ignorant (Maranta et al. 2003). Policy makers and campaigners who are developing strategies to 'educate' consuming publics about better practices of buying and eating food imagine a 'virtual' consumer who is unknowing or even misled (e.g. Freidberg 2004; Hughes 2004; Morris and Young 2004).

Imagining environmental publics as unknowing, as distanced from production and its horrors, as these arguments do, leads to a search for solutions to the 'problem'. Researchers have argued that we should attempt to reverse this clouding of the conditions of production by reemphasising the unappetizing materialities of production, whether this be the treatment of chickens or of sweatshop workers (e.g. Hartwick 2000; Hudson and Hudson 2003; Schlosser 2002; Watts 2004). Researchers in political economy and rural geography have urged the deliberate, "pedagogical" (Hudson and Hudson 2003, p. 427), re-education of consumers to shift them towards better practices through a "politics of reconnection" (Hartwick 1998, p. 433) that aims to change consumers' imaginaries of food and other commodities, raising their awareness of links between consumption and production locally and globally (e.g. Eden et al. 2008; Goodman 2004; Lockie 2009; Morris and Kirwan 2010). This "politics of reconnection" can take two different forms: a politics of local reconnection and a politics of connection through alterity. Let us look in more detail at those in turn.

Politics of local reconnection

The politics of local reconnection advocates shorter supply chains that bring consumers face-to-face with producers through farmers' markets, farmgate sales and organic box schemes (e.g. Ilbery et al. 2005; Holloway and Kneafsey 2000; Seyfang 2006; Watts et al. 2005; for a critique, see Hinrichs 2000). These encourage consumers to buy locally produced goods and/or in local (i.e. small) shopping systems, so that consumers encounter producers directly through the 'spatial fix' of moving consumption closer to production.

And as well as the spatial fix of bringing consumers and producers together in farm shops or markets, an 'information fix' is encouraged to promote traceability and connection with production through labelling, e.g. Sustain's 'Sustainable Food Chains' (analysed by Jackson et al. 2006), Marks and Spencer's 'named farmer' scheme or the Yorkshire Soup Company's 'local heroes' products that use photographs and biographical details of 'real' farmers on packaging to sell produce through both national and regional connections. In a similar way, Figure 4.1 shows a label from a box of mushrooms sold in a shop 2 miles from where I live, which have been grown in another village a few miles away from the same shop.

There are several problems with such attempts to reconnect consumers and producers. First, 'local' arguments can support 'defensive localism' or nationalism, e.g. 'Buy British' arguments, as much as environmental or ethical consumption practices. Associating the 'local' with other non-geographical attributes is thus highly suspect and emphasises the geographical fetishism that conceives of distance as the problem and (especially local) reconnection as the solution (e.g. Holloway and Kneafsey 2000; Ilbery and Kneafsey 1999; Ilbery and Maye 2006; Princen 1997; Winter 2003).

Second, consuming involves interacting with producers especially through in-person 'facework' (Giddens 1990) when producers 'meet and greet' their consumers directly at market stalls or farm shops. Reconnecting through consuming

FIGURE 4.1   Mushrooms labelled with local grower's address (photograph by the author).

practices thus requires more social effort in terms of communication skills and time investment, as well as travelling, than online shopping, for instance, and it assumes that consuming publics are able to judge the quality of information from a farmer or producer that they talk to, as well as judge the quality of the produce they are selling.

And buying locally may involve reading about and seeking out relevant products and retailers, online or through personal networks and recommendations or publicity, such as 'local food awards' covered in the regional or national press. Again, consuming practices include practices of learning to consume, which again shape the consumer. Reading Schlosser's (2002) indictment of the agribusiness supporting fast food production in the USA or talking to a farmer about how chickens suffer in battery farming may involve consumer learning and change, e.g. becoming a vegetarian or reducing their consumption of particular kinds of foods. So learning shapes consuming, even as consuming shapes publics.

Third, this agenda of local reconnection risks romanticising and over-valorising 'local' food, which can limit consumer imaginations, narrowing their scope and obscuring other, more explicitly politicised consuming practices:

> the promotion of localism as the only logical alternative to global capitalism circumscribes the scope and scale of active citizenship in more-or-less exactly the same way as do neoliberal models of the self-monitoring and self-regulating citizen-consumer/entrepreneur.
>
> *Lockie 2009, p. 200*

Politics of reconnection through alternativeness

Another way of advocating reconnection is through 'alternative' supply chains, which are set up in opposition to environmentally damaging and socially exploitative 'mainstream' production. For example, rather than buy local food, consumers can buy food which is certified by independent parties as produced in better (more ethical, more environmental) ways. Such food may be local as well, e.g. I can buy organic vegetables grown in Yorkshire in a local shop in Yorkshire, although they may be certified as organically grown by a national organisation such as the Soil Association. But often certification is applied within global supply chains, e.g. FairTrade products, food certified as organically grown overseas or timber and paper certified by the Forest Stewardship Council as sustainably produced.

The argument is that such practices can promote traceability through a 'knowledge fix' of assurance schemes and labelling (e.g. Eden et al. 2008a, 2008b), so that even where producers and consumers do not encounter each other in person, consumers can still trust the supply chain to be provide an environmentally and/or ethically better 'alternative' to the mainstream (e.g. Morris and Young, 2004). Product assurance schemes seek to guarantee qualities of the product that the consumer cannot themselves detect or evaluate, such as whether an item was organically grown or manufactured without exploiting sweatshop labour – what are sometimes called 'credence' or 'proxy' qualities. Rather than face-to-face contact as discussed above, this relies on product labelling, online directories or social media to provide information and reassurance. Such schemes often use 'knowledge intermediaries', from famous names to charities, to assure consumers that the information they provide is valid and actionable: celebrity chefs are often high-profile champions of food sustainability and welfare standards in the UK, for instance.

Traceability has also been incorporated into the 'Internet of Things' where physical objects (e.g. products, animals) are embedded into digital networks, through which they can communicate with different users. Putting digitally readable information directly onto food products enables consumers to scan QR codes, scan barcodes or type codes into a website in order to learn more about the producers and production process at the moment of purchase in a shop. For example, I bought a box of six eggs in a farm shop near York and each bore a serial number and a tiny lion symbol marked in red ink on the eggshell (see Figure 4.2). Following instructions on the egg box, I went to the website for the egg traceability system run by the British Lion mark (see Figure 4.3), typed in the serial number and (in a couple of clicks) was given the address of the farm where the egg was laid and the farmer's name.

Even the materiality of food, therefore, is exploited for digital politics of traceability in a world of multiple virtual presences, to connect producers and consumers and engender trust. This echoes arguments that online practices can help to democratise consumption (e.g. Denegri-Knott & Molesworth 2010; Ritzer 1999),

FIGURE 4.2 Codes on eggs for digital traceability (photograph by the author).

FIGURE 4.3 Screenshot of digital traceability website for eggs (Used by permission from the British Egg Industry Council).

by distributing responsibility more interactively between networked consumers and producers and shifting practices towards sustainability.

But the knowledge fix of certification and assurance to consumers remains problematic because information does not flow in one direction to a passive recipient as may be assumed; instead, information about products can be re-interpreted, validated, received, resisted and outright ignored as people practise consumption. Even when telling 'food stories' online or through scanning an egg in the Internet of Things, not all information is available or trustworthy; multiple exclusions, silences and virtual absences continue to frame these stories. And again, accessing digital information through a smartphone or computer bundles together different consuming practices of energy, food and information.

And knowledge/information does not necessarily have the desired effect on environmental publics anyway. As we saw in Chapter 2, studies show that providing information (even in a neutral form) can have negative effects on people's attitudes towards new technologies such as genetic modification (e.g. Poortinga and Pidgeon 2004; Scholderer and Frewer 2003). In a survey for Defra (Thornton 2009, p. 57), when respondents were asked how they would change their practices "if they had a better understanding of the environmental impacts of how food is produced", only half of those surveyed said that they would be willing to change what food they bought to reduce their environmental impact, with another 23 per cent saying they would continue to buy the same food. And this is only what they were willing to say in a survey, not what they actually did, which is even less likely to be influenced by 'a better understanding'.

At a more structural level, consumption practices are also constrained by income, knowledge, personal circumstances, systems of provision and social norms. There are also 'digital divides' between consumers, with some readily using the internet for all sorts of purposes, including consuming, and others not able or willing to because of age, lack of interest or lack of broadband or mobile reception.

So, consuming information about the certified origins of sustainable or ethical products is a complex process of learning, decoding and giving meaning; imaginaries of both consumers and commodities are not givens, but are produced through practice in ways that are continually changing and spatialized through multiple geographies. Knowing and consuming are thus related, but not causally: sometimes, consuming itself is a means of gathering knowledge, e.g. through reading labelling; sometimes, what is consumed is itself knowledge, e.g. through buying consumer guide books or magazines. But information is usually seen as a public good and rarely rejected by public opinion: when asked, people rarely say that would prefer to be given *less* information, but it does not follow that, if they are given *more* information, they will change their behaviour as a consequence of that learning.

Power

In this final section, we come to probably the most contentious aspect of consuming publics: whether and how they are powerful. This question is fraught with controversy and moral loadings of different kinds that crop up not only in academic debates, but also in everyday conversations, politics and perceptions of the rights and duties of publics in different settings. Casting environmental publics as primarily consumers is a very common way for commentators to undermine the power of consumption practices, even as other commentators celebrate it, so let us explore this contradiction by considering how practices are theorised differently.

Consuming as good practice

To take the positive theorisation first, the phrase 'consumerism' can be used to imply that consumers are a force for change in the world, able to support good causes and put pressure on producers. In times of economic recession, governments often try to promote spending by consumers, arguing that buying things makes businesses grow and is thus patriotic in supporting national economic growth. In the late 19th century, "consumer interest became the public interest" (Soper and Trentman 2008, p. 6) and in the 20th century, 'digging for victory' and recycling rubber and metal for military use in times of scarcity during World Wars were portrayed as the virtuous acts of patriotic consumers.

More recently, during the post-2008 global recession, consumption has entered public interest again as people are urged by national governments to spend their money and buy products made in their own country to support economic growth, making consumption not merely a public duty but an expression of patriotism, and practices of consumption also practices of citizenship. In the UK, the slogan 'Buy British' illustrates this, but so does the barrage of news stories in and around Christmas every year, speculating how much consumers will spend and therefore whether it will be a 'good Christmas' for retailers (and for the country) or not.

In a very different way, consumerism is seen as empowering ordinary consumers to use their consuming practices to challenge producers and retailers and resist their dominance through boycotting products and raising awareness of alternative products. Examples include the 'slow food' movement, campaigns against proposals to build big supermarkets that might threaten local shops and campaigns for more sustainable consumption such as promoting food certified as organically grown or sustainably produced, low-energy light bulbs and recycled toilet paper. Also called 'political consumerism', such harnessing of consumer power to perform politics through every day means has a long history over the 20th century, from Mahatma Gandhi urging Indians to wear homespun cotton rather than buy cotton imported from England during India's fight for independence, to anti-apartheid campaigns urging consumers across the world to boycott produce from South Africa.

Later, the best-selling book, *The Green Consumer Guide* (Elkington and Hailes 1990), was credited by some with kicking off interest in green consumption in the UK in the 1990s, identifying as it did the possibilities for a green market in everyday items from laundry detergents to paints. Its authors regarded the politics of consumerism and reconnection very positively, echoing the arguments of many others who have regarded "consumption as empowerment" (Miller 1995, p. 41) and "may be the most important force that unites the contemporary world" (Firat & Dholakia 1998, p. 103).

Consumption is perhaps most effective when individuals consume together: cooperatives and group-purchasing by individuals is today being facilitated by the internet more than ever. This can reduce costs. For example, when twenty households together place an order for oil, they can get cheaper prices because of an order of such a large size is more attractive to the supplier, or lever social change by refocussing consumption from individual desire to the public good. For example, Carbon Reduction Action Groups (CRAGs) are voluntary groups of individuals who are trying to reduce their personal carbon footprint, that is, de-consuming carbon. CRAGs have been interpreted as expressions of civic governance, public environmentalism and ecological citizenship, rather than of self-centred consumption, precisely because of this supposed sacrifice to the greater public good of averting climate change (e.g. Hulme 2009, p. 307). This incarnates consuming as good environmental practice and as enabling individual and societal change.

Consuming as bad practice

But when negatively constructed, 'consumerism' implies self-interested practices in which gratification of desire through spending money and having stuff is all-important. It also imagines consumers not as active agents of their own practices, but as dupes manipulated by worldwide systems of advertising and retailing to buy stuff that they do not need in order to feel good and to feel part of society. Frequent in analyses inspired by political economy, this construction portrays a 'green' market not as reflecting consumer attitudes, but deliberately created by capitalism to generate more opportunities for accumulating profit and expansion, usually to the detriment of the environment and the health and wellbeing of (most of) humanity. In such analyses, "consumption has throughout history been seen as intrinsically evil" (Miller 2001, pp. 226–227), because it serves capitalism, rather than the public good.

Such analyses locate power in the capitalist system and especially with producers, imagining consumers as largely powerless and alienated from industrialised systems of production, either because those systems are invisible to everyday consumers or because workers in those systems (who are also consumers) are not inspired by or skilled in their work, but are rather treated like cogs in a machine – paid but not participating in how production is designed and operated. In such analyses, any campaigns for ethical and sustainable consumption are seen as merely delusions on the part of consumers, and the commodification of ethics also turns ethical and environmental values into more products to be advertised and promoted for the benefit of capitalism.

Because of such interpretations, supporters of 'alternative' food networks often become suspicious when those networks succeed and grow (e.g. Goodman 2000; Guthman 2003, 2004; Mansfield 2004; Raynolds 2000), doubting that large-scale production of organic food can match the same environmental and social standards as 'alternative' (i.e. small-scale) production, and worrying that growth will turn the 'alternative' into what it opposes. For such commentators, small is beautiful and large is nasty when it comes to moralising about ethical retail, regardless of the potential benefit to the environment of large-scale shifts in production and consumption toward less damaging production, whether this is organic production becoming agribusiness, (e.g. Guthman 2004), Nestlé certifying KitKat bars as FairTrade or Starbucks (2014) sourcing "socially responsible" coffee through Conservation International.

This has led to arguments that Fairtrade and other ethical labels are now merely brands, no longer alternative nor challenging the environmental and social exploitation in global production by multinational corporations. Some argue that consuming such products now represents not caring or compassion but fashion, with celebrities used to advertise ethical products and charities to make them more attractive in ways that their ethical credentials perhaps cannot (see Goodman 2010). Consumers are perceived to be naïve in following famous names but also selfish in consuming ethical products not because of their ethics but because of their associated glamour.

Moreover, this roll-out of niche practices of producing, retailing and consuming to the mainstream, what has been called the 'conventionalisation' of supposedly 'alternative' production (Guthman 2004), means that ethical values can be enacted by more and different types of consumer (e.g. Lockie 2009), not only those who buy in small shops and through cooperatives, but also those who buy organic produce in supermarkets, which consequently have the biggest share of the organic market: 70 per cent of the £1.86 billion worth of UK organically certified produce was sold through chain stores in 2014 (Soil Association 2015).

All these arguments produce a moral ordering of consuming as less worthy than other practices also performed by people who consume, such as voting (Chapter 7) or participating in democratic debates about environmental decisions to be made by the state (Chapter 3). This ordering can then be transferred from practices to publics, to portray consumers as acquisitive, materialistic, superficial, individualised, wasteful, inauthentic and unpolitical, but citizens as morally worthy, politically aware, concerned with the public good and empowered by that to seek change. The two imagined figures of 'consumer' and 'citizen' are thus dualistically opposed:

> individuals *qua* consumers have most often been presented as obedient to forms of self-interest that either limit or altogether preclude the capacity for the reflexivity, social accountability and cultural community associated with citizenship. Only in their role as citizens do they supposedly look above the parapet of private needs and desires or could be said to have an eye to the public good. This perception is further reinforced in the theoretical division between a public domain of citizenship – and its concerns with rights, duties, participation and equality – and the private domain of the supposedly purely self-interested consumer.
>
> *Soper 2007, p. 206*

As Soper and Trentman (2008, p. 4) noted, this dualism of consumer versus citizen is restrictive, because it unrealistically separates consuming from other, more citizenly practices, and makes the consumer the villain and the citizen the saviour: "the consumer is conceived as the cause of the unfortunate environmental consequences that the citizen as active campaigner will attempt to correct or alleviate" (Soper 2007, p. 208). But an individual is both: I may buy some lunch on my way to participate in a public inquiry about a new development, for instance, so performing both as a 'consumer' and a 'citizen' at different times and in different places. The imagined figures are thus exercises in moral and political valuing that hamper our understanding of how practices of consuming are enacted and given meaning and of how practices interact through people's everyday lives.

Soper (2007, p. 207) argued that blending these two figures into a single figure of the 'consumer-citizen' would be more useful, because it would fuse rather than distinguish between a person's self-interest and their care for others, and re-politicize consumption. Analyses often do conflate consumers and citizens, sometimes implicitly even as authors explicitly ascribe different values to the practices that both consumers and citizens enact. For example, Lockie (2009, p. 200) argued that "'food citizens' must act reflexively and proactively to re-invent for themselves their identities and practices as food consumers", and Seyfang (2005, p. 292) that "sustainable consumption is clearly identified as a tool for practising ecological citizenship – requiring individuals to make political and environmental choices in their private consumption decisions".

To move away from this persistent dichotomy of consumer versus citizen, Soper (2007) suggested that conceiving of ethical consuming as 'alternative hedonism' would enable us to see consuming practices also as citizenly, that is, to see that people may buy things while being self-interested as well as altruistic: people may derive pleasure from feeling that they are avoiding environmental damage or worker exploitation by buying something. So instead of green consumption being a hair-shirted lifestyle of restrictions and no fun, it can be a positive experience that combines "the sensual pleasures and the moral rewards of consuming sustainably" (Soper 2007, p. 212). For example, the public good of reducing carbon emissions may be as important as personally avoiding the stress of negotiating traffic when a consumer decides to travel to work by train instead of by car.

The problem is that the same practice is differently moralised: if an action is denigrated as (merely) consuming, it generates less support from some quarters than if it is presented as citizen action, and can restrict how well other groups engage with environmental publics.

> The supposed blindness of consumers to anything but their private interests is thereupon invoked as grounds for 'experts' to profess their scepticism about the wisdom of any form of consultation exercise.
>
> *Soper 2007, p. 219*

This fits very well into the arguments made so far in this book, especially that practices (rather than individuals) are what matter and are what we need to analyse. Similarly, Spaargaren (2011) argued that sustainable consumption research tends to adopt either an individualistic paradigm of consuming that is associated with psychology and dominated by individual agency, putting too much emphasis on voluntary changes in the interests of the environment, or a systemic paradigm that is associated with sociology and dominated by structural conditions that vastly restrict an individual's power to effect change. By focussing on practices, their geographies and implications for power, rather on individuals or structures, we can perhaps get beyond this unhelpful binary of consumer-citizen.

Indeed, Soper argued that "alternative hedonism" may be stronger and more effective in promoting sustainable consumption because serving the public good can also be self-gratifying: "where material self-interest acts as an additional and complementary motive for more environmentally friendly practice, its impact over time is likely to be considerably greater" (Soper 2007, p. 221). This argument thus aims to represent ethical and sustainable consumption not as giving up things or the foregoing pleasures of buying and using things, but to represent consuming as also caring and enjoying caring. This also echoes Miller's (1998) argument in *A Theory of Shopping* that people consume for love, whether buying birthday presents, the latest book by a partner's favourite author, or a child's favourite flavour of ice cream. Such arguments about consuming being about other people rather than about the self make it even more explicit that consuming practices are social practices, given meaning through interaction, imagination and expectation.

Summing up

Consuming practices are highly diverse, not least because nearly everybody consumes and thus has some environmental impact, and that includes not only what we buy, but also how we use the things that we buy and how we dispose of the residue when we are finished consuming them. Consumers are also often seen as unknowing, echoing the deficit model in Chapter 2, as well as disempowered by consuming within a capitalist system that both hides environmental damage and promotes 'green' brands as easy shifts towards sustainable and guilt-free consumption.

Unknowing is also spatialized within a politics of disconnection, where consumers are seen as increasingly distanced from production, both geographically and topologically, and thus less able to understand and judge products in terms of their likely ethical or environmental qualities. Various proxy qualities have been offered to address this in a politics of reconnection, principally through local and embodied encounters between producers and consumers (a spatial fix) and through alternative methods of production that can be certified as more ethical and/or sustainable by intermediaries (a knowledge fix), although both remain problematic.

In terms of power, analysing environmental publics as consumers continues to be riven by a perceived dichotomy between consumers and citizens, between the private sphere and the public sphere, between selfish, materialistic gratification and democratic, politicised engagement in society. This dichotomy denies consumers power and influence and even denigrates them as dupes of advertising and fashion in ways that ignore how they productively perform both themselves and their practices. I argue that consuming practices are what matter and are what we need to analyse, rather than individuals as consumers, so that we can better understand how environmental publics are performed, made and changed.

Notes

1   This may be much higher when they are at work (see Chapter 8).
2   Some Freecycle groups have changed into Freegle groups, with different online hosting arrangements, but their principles and practices remain very similar.

References

Barnett, Clive, Paul Cloke, Nick Clarke and Alice Malpass (2010). *Globalizing Responsibility: the political rationalities of ethical consumption.* Wiley-Blackwell, Chichester.

BBC (2007). It's in the bag, darling. http://news.bbc.co.uk/1/hi/magazine/6587169.stm

BBC (2015). *Volkswagen: the scandal explained.* www.bbc.co.uk/news/business-34324772

Clapp, Jennifer and Swanston, Linda (2009). Doing away with plastic shopping bags: international patterns of norm emergence and policy implementation. *Environmental Politics* 18, 3, 315–332.

The Cooperative (2012). Ethical consumer markets report 2012. www.co-operative.coop/ P.Files/416561607/Ethical-Consumer-Markets-Report-2012.pdf

Crang, Mike, Alex Hughes, Nicky Gregson, Lucy Norris and Farid Ahamed (2012). Rethinking governance and value in commodity chains through global recycling networks. *Transactions of the Institute of British Geographers* 38, 12–24.

Currah, Andrew (2003). The virtual geographies of retail display. *Journal of Consumer Culture* 3, 1, 5–37.

Denegri-Knott, Janice and Mike Molesworth (2010). Concepts and practices of digital virtual consumption. *Consumption Markets & Culture* 13, 2, 109–132.

Department of Energy and Climate Change (2012). Energy trends. Department of Energy and Climate Change, London. www.gov.uk/government/uploads/system/uploads/attachment_data/file/65906/7343-energy-trends-december-2D012.pdf

Department of Health (2015). 2010 to 2015 government policy: obesity and healthy eating. Department of Health, London. www.gov.uk/government/publications/2010-to-2015-government-policy-obesity-and-healthy-eating/2010-to-2015-government-policy-obesity-and-healthy-eating

Duffy, Rachel, Andrew Fearne and Victoria Healing (2005). Reconnection in the UK food chain: bridging the communication gap between food producers and consumers. *British Food Journal* 107, 1, 17–33.

Eden, Sally, Christopher Bear and Gordon Walker (2008a). Understanding and (dis)trusting food assurance schemes: consumer confidence and the "knowledge fix". *Journal of Rural Studies* 24, 1–14.

Eden, Sally, Christopher Bear and Gordon Walker (2008b). Mucky carrots and other proxies: problematising the knowledge-fix for sustainable and ethical consumption. *Geoforum* 39, 1044–1057.

Elkington, John and Julia Hailes (1988). *The Green Consumer Guide*. Victor Gollancz, London.

Ethical Consumer Research Association (2015). Ethical consumer markets report 2015. www.ethicalconsumer.org/researchhub/ukethicalmarket.aspx

Firat, A. Fuat and Nikhilesh Dholakia (1998). *Consuming People: from political economy to theaters of consumption*. Routledge, London.

Freidberg, Susanne (2004). The ethical complex of corporate food power. *Environment and Planning D: Society and Space* 22, 513–531.

Giddens, Anthony (1990). *The Consequences of Modernity*. Polity Press, Cambridge.

Goodman, David (2000). Organic and conventional agriculture: materializing discourse and agro-ecological managerialism. *Agriculture and Human Values* 17, 215–219.

Goodman, Michael K. (2004). Reading fair trade: political ecological imaginary and the moral economy of fair trade foods. *Political Geography* 23, 891–915.

Gram-Hanssen, Kirsten (2011). Understanding change and continuity in residential energy consumption. *Journal of Consumer Culture* 11, 61–78.

Gregory, Jane and Steve Miller (1998). *Science in Public: communication, culture, and credibility*. Plenum Trade Press, New York.

Gregson, Nicky, Alan Metcalfe and Louise Crewe (2007). Moving things along: the conduits and practices of divestment in consumption. *Transactions of the Institute of British Geographers* 32, 2, 187–200.

Guthman, Julie (2003). Fast food/organic food: reflexive tastes and the making of "yuppie chow". *Social and Cultural Geography* 4, 1, 45–58.

Guthman, Julie (2004). Back to the land: the paradox of organic food standards. *Environment and Planning A* 36, 3, 511–528.

Hargreaves, Tom, Michael Nye and Jacquelin Burgess (2010). Making energy visible: a qualitative field study of how householders interact with feedback from smart energy monitors. *Energy Policy* 38, 6111–6119.

Hargreaves, Tom, Noel Longhurst and Gill Seyfang (2013). Up, down, round and round: connecting regimes and practices in innovation for sustainability. *Environment and Planning A* 45, 402–420.

Harvey, David (1990). Between space and time: reflections on the geographical imagination. *Annals of the Association of American Geographers* 80, 418–434.

Hartwick, Elaine (1998). Geographies of consumption: a commodity-chain approach. *Environment and Planning D: Society and Space* 16, 423–437.

Hartwick, Elaine R. (2000). Towards a geographical politics of consumption. *Environment and Planning A* 32, 1177–1192.

Hinrichs, Clare (2000). Embeddedness and local food systems: notes on two types of direct agricultural market. *Journal of Rural Studies* 16, 295–329.

Holloway, Lewis and Moya Kneafsey (2000). Reading the space of the farmers' market: a preliminary investigation from the UK. *Sociologia Ruralis* 40, 3, 285–299.

Hoppner, Corrina and Lorraine Whitmarsh (2011). Public engagement in climate action: policy and public expectations. 47–65 in Lorraine Whitmarsh, Saffron O'Neill and Irene Lorenzoni (edited), *Engaging the Public with Climate Change: behaviour change and communication*. Earthscan, London.

Hudson, Ian and Mark Hudson (2003). Removing the veil? *Organization & Environment* 16, 4, 413–430.

Hughes, Al (2004). Accounting for ethical trade: global commodity networks, virtualism and the audit economy. In Hughes, Al and Susy Reimer (editor), *Geographies of Commodity Chains*, London: Routledge, 215–232.

Hulme, Mike (2009). *Why We Disagree about Climate Change*. Cambridge University Press, Cambridge.

Ilbery, B. and M. Kneafsey (1999). Niche markets and regional speciality food products in Europe: towards a research agenda. *Environment and Planning A* 31, 2207–2222.

Ilbery, B. and D. Maye (2006). Retailing local food in the Scottish-English borders: a supply chain perspective. *Geoforum* 37, 352–367.

Ilbery, Brian, Carol Morris, Henry Buller, Damian Maye and Moya Kneafsey (2005). Product, process and place: an examination of food marketing and labelling schemes in Europe and North America. *European Urban and Regional Studies* 12, 2, 116–132.

Jackson, Peter, Neil Ward and Polly Russell (2006). Mobilising the commodity chain concept in the politics of food and farming. *Journal of Rural Studies* 22, 129–141.

Lockie, Stewart (2009). Responsibility and agency within alternative food networks: assembling the "citizen consumer". *Agriculture & Human Values* 26, 193–201.

Mansfield, Becky (2004). Organic views of nature: the debate over organic certification for aquatic animals. *Sociologia Ruralis* 44, 2, 216–232.

Maranta, Alessandro, Michael Guggenheim, Priska Gisler and Christian Pohl (2003). The reality of experts and the imagined lay person. *Acta Sociologica* 46, 2, 150–65.

Micheletti, Michele and Dietlind Stolle (2008). Fashioning social justice through political consumerism, capitalism, and the internet. *Cultural Studies* 22, 5, 749–769.

Miller, Daniel (1995). *Acknowledging Consumption*. Routledge, London.

Miller, Daniel (1998). *A Theory of Shopping*. Polity Press.

Miller, Daniel (2001). The poverty of morality. *Journal of Consumer Culture* 1, 2, 225–243.

Morris, Carol and Craig Young (2004). New geographies of agro-food chains: an analysis of UK quality assurance schemes. 83–101 in Alex Hughes and Suzanne Reimer (edited), *Geographies of Commodity Chains*. Routledge, London.

Morris, Carol and James Kirwan (2010). Food commodities, geographical knowledges and the reconnection of production and consumption: the case of naturally embedded food products. *Geoforum* 41, 131–143.

Naylor, Lindsay (2012). Hired gardens and the question of transgression: lawns, food gardens and the business of "alternative" food practice. *Cultural Geographies* 2012 19, 483–504.

Ozaki, Ritsuko, Isabel Shaw and Mark Dodgson (2013). The coproduction of "sustainability": negotiated practices and the Prius. *Science, Technology, & Human Values* 38, 518–541.

Poortinga, Wouter and Nick F. Pidgeon (2004). Trust, the asymmetry principle, and the role of prior beliefs. *Risk Analysis* 24, 6, 1475–1486.

Princen, Thomas (1997). The shading and distancing of commerce: when internalization is not enough. *Ecological Economics* 20, 235–253.

Raynolds, Laura T. (2000). Re-embedding global agriculture: the international organic and fair trade movements. *Agriculture and Human Values* 17, 297–209.

Ritzer, George (1999). *Enchanting a Disenchanted World.* Pine Forge Press, Thousand Oaks, CA.

Ritzer, George and Nathan Jurgenson (2010). Production, consumption, prosumption: the nature of capitalism in the age of the digital "prosumer". *Journal of Consumer Culture* 10, 13–36

Rogers, Richard (2012). Mapping and the politics of web space. *Theory, Culture & Society* 29, 4/5, 193–219.

Sayer, Andrew (2003). (De)commodification, consumer culture, and moral economy. *Environment and Planning D: Society and Space* 21, 341–357.

Schlosser, Eric (2002). *Fast Food Nation: what the all-American meal is doing to the world.* Penguin, London.

Scholderer, Joachim and Lynn J. Frewer (2003). The biotechnology communication paradox: experimental evidence and the need for a new strategy. *Journal of Consumer Policy* 26, 125–157.

Seyfang, Gill (2005). Shopping for sustainability: can sustainable consumption promote ecological citizenship? *Environmental Politics* 14, 2, 290–306.

Seyfang, Gill (2006). Ecological citizenship and sustainable consumption: examining local organic food networks. *Journal of Rural Studies* 22, 4, 383–395.

Shove, Elizabeth (2003). *Comfort, Cleanliness and Convenience.* Berg.

Shove, Elizabeth (2010). Beyond the ABC: climate change policy and theories of social change. *Environment & Planning A* 42, 1273–1285.

Soil Association (2015). *UK Organic Market 2015.* Soil Association, Bristol. https://secure-payment.soilassociation.org/p./contribute/organicmarketreport2015

Soper, Kate (2007). Re-thinking the "good life": the citizenship dimension of consumer disaffection with consumerism. *Journal of Consumer Culture* 7, 2, 205–229.

Soper, Kate and Frank Trentman (2008). Introduction. 1–16 in Kate Soper and Frank Trentman (edited), *Citizenship and Consumption.* Palgrave Macmillan, Basingstoke.

Spaargaren, Gert (2011). Theories of practices: agency, technology, and culture. Exploring the relevance of practice theories for the governance of sustainable consumption practices in the new world-order. *Global Environmental Change* 21, 813–822.

Star, Susan Leigh (1990). Power, technology and the phenomenology of conventions: on being allergic to onions. *The Sociological Review* 38, S1, 26–56.

Starbucks (2014). Responsibly grown coffee. http://starbucks.co.uk/responsibility/sourcing/coffee

Thornton, Alex (2009). Public attitudes and behaviours towards the environment – tracker survey: a report to the department for Environment, Food and Rural Affairs.TNS. Defra, London. https://data.gov.uk/dataset/survey_of_public_attitudes_and_behaviours_towards_the_environment/resource/485d7556-0a6d-4d01-8b82-9795b6cbd009

TNS (2011). Eurobarometer 365: Attitudes of European citizens towards the environment. TNS Opinion and Social, Brussels.

Volkswagen Group UK (2015). Volkswagen UK announces action plan to modify diesel vehicles with EA 189 EU5 engines. www.seat.co.uk/owners/diesel-engines/statement/statement-volkswagen-uk.html

Warde, Alan (2005). Consumption and theories of practice. *Journal of Consumer Culture* 5, 2, 131–153.

Warde, Alan (2014). After taste: culture, consumption and theories of practice. *Journal of Consumer Culture* 14, 279–303.

Watts, D.C.H., B. Ilbery and D. Maye (2005). Making reconnections in agro-food geography: alternative systems of food provision. *Progress in Human Geography* 29, 1, 22–40.

Watts, Michael J. (2004). Are hogs like chickens? Enclosure and mechanization in two "white meat" filières. 39–62 in Alex Hughes and Suzanne Reimer (edited), *Geographies of Commodity Chains*. Routledge, London.

Welsh Government (2014). Single-use carrier bags. http://gov.wales/topics/environment-countryside/epq/waste_recycling/substance/carrierbags/?lang=en

Winter, Michael (2003). Embeddedness, the new food economy and defensive localism. *Journal of Rural Studies* 19, 23–32.

WRAP (2010). Environmental Life Cycle Assessment (LCA) study of replacement and refurbishment options for domestic washing machines. WRAP, London. www.wrap.org.uk/sites/files/wrap/Washing_machine_summary_report.pdf

WRAP (2012). *Household Food and Drink Waste in the United Kingdom 2012*. WRAP, London. www.wrap.org.uk/content/household-food-and-drink-waste-uk-2012

WRAP (2015). How much food is wasted in total across the UK? WRAP, London. www.lovefoodhatewaste.com/content/how-much-food-wasted-total-across-uk

Zwick, Detlev, Samuel K Bonsu and Aron Darmody (2008). Putting consumers to work: "co-creation" and new marketing govern-mentality. *Journal of Consumer Culture* 8, 163–197.

5
ENJOYING PUBLICS

Introduction

In 2011, over 10,000 people donned wetsuits and plunged into the waters of Lake Windermere in the English Lake District as part of the Great North Swim, sponsored by British Gas. All of the places available for the competitive long-distance swim were taken within six weeks of the launch (Great Swim 2011), despite (or perhaps because of) the cancellation of the event the previous year, when the Environment Agency found blue-green algae in the lake water, algae which could be dangerous to human health by causing sickness or skin rashes (BBC 2010).

Later that same year, David Walliams, actor and star of the UK television comedy show *Little Britain*, swam 140 miles of the river Thames to raise money for charity as part of Sport Relief. Reportedly due to Thames Water dumping 500,000 m³ of sewage into a section he was about to swim, Walliams suffered 'Thames tummy' in the shape of diarrhoea and vomiting, as well as making the front pages of national newspapers (e.g. *The Guardian* 2011).

Why do so many people go swimming (or watch people swimming) as a way of enjoying the environment, especially when it might involve getting cold, wet, tired and ill? Increasing coverage of outdoor swimming on television is one of many indications of how the environment is used as a playground by people, despite many claims that we are increasingly an indoor or sedentary society. These two examples also emphasise that the environment is a risky playground, especially for those playing in possibly polluted waters.

This chapter is about how the environment is used for fun and what this means for the publics that are produced through this sort of environmental engagement. Seeing the environment as a playground is a relatively new idea in human history; in the past, the environment was more usually seen as a source of danger from wild animals and weather events, as well as a place of toil in the form of hard labour in agricultural fields, with towns and cities being oases of relative peace by comparison. Take Cumbria in northern England, now a very popular holiday destination because of the lovely Lake District scenery, but in the 1720s known as Westmoreland and described by the novelist Daniel Defoe as "a country eminent only for being the wildest, most barren and frightful of any that I have passed over in England, or even in Wales".

But with the rise of the Romantic movement in art and literature and its veneration for natural beauty in the 19th century and then the rise of environmental recreation through to the early 20th century, perspectives began to shift, giving more value to environmental enjoyment and exercise. Environmental recreation became popular at first mainly among the elite and wealthy, but later spread through all classes, with bicycling, hiking and outdoor exercise being popularised in the 1930s and new sports and activities continually being invented through the 20th century, such as windsurfing.

Today, not only do many people play outdoors, but they also party outdoors, picnicking, barbecuing and even scuba-diving to get married underwater. To analyse how such recreations shape environmental publics and their role in environmental management, this chapter will consider the highly diverse practices involved in environmental recreation today, their geographies and their implications for the power of environmental publics.

One point to note immediately is that, unlike previous chapters in which practices have environmental consequences that are often far distant from where the practices take place (such as food consumed in an English household that has been grown on a Kenyan farm), in the case of environmental recreation, environmental publics are frequently defined and performed often in the very environment that they are enjoying, that is, through *in situ* environmental practices. Another key point is that, although all human activities evolved outdoors, today, outdoor recreation is often only a minority interest, with the result that the environmental publics that we shall meet in this chapter are much more specialised and fewer in number than those we have met so far in this book.

Practices of environmental recreation

Let us start with that point, by considering numbers. Walking is the most common outdoor recreation reported by people in the UK (e.g. Natural England 2014) and the importance of urban parks for local residents has been a mainstay of planning and research for over a century, with Victorian philanthropy idealising outdoor recreation as healthy for mind and body. In 2011, 56 per cent (63 per cent of ABs, 47 per cent of DEs) of people surveyed for Defra reported visiting green spaces such as parks or commons at least once a week and 13 per cent reported volunteering for or participating in conservation work at least once a year (Defra 2011).

Such everyday recreation is a very good example of the embodied practice of environmental engagement by a wide variety of publics, both old and young, low and high income, and is frequently shared with children or other family

TABLE 5.1 Most popular reasons for outdoor recreation away from home in the UK, ranked

Rank	Reasons	% surveyed people who reported this	Estimated number of visits annually
1.	Walking with a dog	50%	1,474 million
2.	Walking, not with a dog	26%	775 million
3.	Playing with children	9%	265 million
4.	Eating or drinking out	5%	159 million
5.	Running	4%	109 million
6.	Visiting an attraction	3%	89 million
7.	Sightsee, picnic, drive	3%	77 million
8.	Informal games and sport	3%	75 million
9.	Road cycling	2%	73 million
10.	Wildlife watching	2%	66 million
11.	Picnicking	2%	49 million
12.	Visits to a beach, sunbathing, paddling	1%	43 million
13.	Off-road cycling, mountain biking	1%	36 million
14.	Appreciating scenery from your car	1%	32 million
15.	Horse riding	1%	30 million
16.	Swimming outdoors	1%	19 million*
17.	Fishing	<1%	14 million
18.	Watersports	<1%	14 million
19.	Fieldsports	<1%	10 million

Source: Redrawn from Natural England (2014, page 22), based on 55,897 respondents.
Note:
★ This estimate doubled between 2012 and 2014.

members, friends and pets. By 'common' or 'everyday', I mean that such practices are undertaken by lots of people and regularly, but not in any highly specialised way, not in specialised places or with any special preparation, but through repetition or routinisation as part of their daily lives.[1] Outdoor recreation may start at the doorstep or involve a little more travel: for example, in the UK, walking (with or without a dog) is by far the most common reason reported by people for travelling away from home to visit 'the natural environment' (see Table 5.1).

As well as going away from home to enjoy the environment, far more people also enjoy it at home, with 67 per cent of adults reporting that they engaged with the natural environment by "sitting or relaxing in my garden" and slightly fewer (53 per cent) did some (more active or embodied) gardening there (Natural England 2014). Many people also reported visually consuming the environment through watching nature programmes on the television or listening to them on the radio, reading nature items in books or websites or watching "natural scenery from indoors or whilst on journeys" (Natural England 2014, see Table 5.2).

The daily routine of such engagement, usually close to home and probably (although Natural England did not ask about this in their survey) in the same places every time, means that not only is environmental engagement enacted through

TABLE 5.2 Estimates of adults participating in less common environmental recreations in England

Recreation	Estimated number of adults over 16 years old in England who participated at least once a week (October 2011–October 2012)	% of population
Angling	129,200	0.30%
Mountaineering	98,700	0.23%
Snowsports	80,400	0.19%
Sailing	64,400	0.15%
Canoeing	46,600	0.11%
Rowing	42,100	0.10%

Source: Mintel Oxygen.

recreation, but family life itself is also performed and continually (re)made through such practices. Involving local people in everyday activities like gardening has also been promoted as a way to improve public participation and social capital (Blomley 2004; Hinchliffe et al. 2007). Likewise, everyday activities provide opportunities for local communities to feel ownership of their local environments, as a more successful way to engage local publics than through abstract policy ideas.

Fewer people participate regularly in less common environmental recreations like surfing, climbing, kayaking and parascending, which have their own specialised equipment, locations, experts and habits. But how should we measure 'participation'? For example, there are around 1.3 million fishing licences sold annually in England and Wales, licences which are legally required for anyone freshwater fishing, but maybe only 250,000 people fish for fun regularly, that is, at least monthly, so different methods produce very different counts of participation. And outdoor recreations that involve very small minorities are notoriously difficult to estimate (see Table 5.3). Some of these more specialised recreations, such as fishing and hunting, are highly dependent upon specific ecologies or environmental resources. In the USA, the US Fish and Wildlife Service (2012) estimated that about 37.4 million people participated in recreational hunting, angling and other wildlife activities in 2011, spending $145 billion (around 1 per cent of GDP).

These practices of environmental recreation, perhaps more than the other practices discussed so far, literally remake the publics that participate in them. Active outdoor recreation builds participants' muscles through regular and demanding exercise, but may also damage their bodies through causing them to fall from heights or to catch water-borne infections that make them ill. So, in bodily terms, environmental recreationists may be physically more or less fit and healthy than non-recreationists, as their bodies are made and remade through the physicalities of leisure. Sharing information about risks, equipment, best practice is therefore important for environmental publics to enjoy outdoor recreation, as well as shaping how they practise it.

TABLE 5.3 Most common activities reported involving the natural environment, ranked

Activity (prompted list)	% surveyed people who reported this
Sitting or relaxing in a garden	67
Choosing to walk through local parks or green spaces on my way to other places	55
Gardening	53
Watching or listening to nature programmes on the TV or radio	52
Looking at natural scenery from indooors or whilst on journeys	44
Watching wildlife	37
Looking at books, photos or websites about the natural world	31
Doing unpaid voluntary work out of doors	7

Source: Redrawn from Natural England (2014, page 29). Question asked: Which of the following activities involving the natural environment do you take part in? 3,535 respondents answered, but some chose more than one activity, so total exceeds 100%.

And recreationists of all kinds also develop routines and accumulate equipment through huge personal investment in specialised technologies, time or rituals, such as endurance/extreme running and 'Munro-bagging' for Scottish mountains, or specialised in other ways, such as Nordic walking using specialised poles (Shove and Pantzar 2005) or rock climbing using nuts and cams (Barratt 2011). Thus are practices of outdoor recreation enacted not solely through the bodily encounter with the environment, but also through the devices and other 'stuff' that enable that encounter and, often, render it more comfortable and safer.

In contrast, a minority of people hardly ever go into the environment to have fun: 21 per cent of people surveyed by Natural England (2014) reported going to parks or other green spaces "less than once a month or never". Low levels of physical activity have been increasingly blamed for poor physical health in urbanised societies, and attempts to combat this in policy have included the promotion of environmental recreation as a prescription for a healthier life, supported by research evidence that outdoor activity promotes mental health (e.g. Hartig and Cooper Marcus 2006; Hartig et al. 2003; Pretty et al. 2007), if not automatically also physical health. For example, the UK government's 'Change for Life' programme has used outdoor exercise to promote healthy lifestyles and family bonding for several years, linking environmental recreation with weight loss and longer life expectancy (see Eden 2009). Natural England, the government's agency for promoting public engagement with the environment, has worked with the Department of Health to promote 'green exercise' for healthy lifestyles and to legitimate environmental protection through linking it with health policies and more secure, long term budgetary commitments.

However, assuming that information about outdoor recreation and physical activity will prompt wider uptake is a common problem in policy that fails to conceptualise how environmental practices develop and spread. Transfer of practices through multiple repeated encounters, often from multiple sources including friends and family and especially through embodied experimentation, is more important than one-way information transfer from experts and policy makers. Prescribing 'green exercise' to build healthier populations suffers from the assumptions of the deficit model that we met in Chapter 2, because information has consistently been shown *not* to change behaviour, despite its appeal to policy makers looking for a quick fix to perceived problems.

Learning and sharing environmental recreational practices

How are environmental publics shaped through outdoor recreation? For specialised practices, learning is enabled through recreational clubs and networks that promote their recreational pursuits, but also often define, legitimate and police what are considered to be acceptable environmental practices for that recreation. This can be done in different ways, such as written codes of conduct and club principles, which vary not only by recreation type (kayaking versus hiking, for instance) but also by space and time. For example, angling clubs have different rules for different stretches of water and their bye-laws as to what can be caught and with what equipment also vary by region and by season. 'Communities of practice' for each recreation are built through informal networks of friends and clubs, but also more formally through central organisations such as the governing body for each sport that is recognised by Sport England (2009), who suggest that 27 per cent of regular anglers and 36 per cent of regular canoeists/kayakers are also a member of a club, receive tuition or take part in a competition each year. So clubs are important in shaping recreational knowledge-practices and their members' environmental engagement, although only a minority of people enjoying each recreation may join a club.

Defining the 'correct' or acceptable forms of environmental recreation normalises and routinizes practice through collective codification. It also allows practices to travel, both through in-person training, where clubs offer workshops with experienced practitioners so that beginners can learn how to perform specialised recreations, or through other, less embodied modes of learning, such as newsletters or website discussion boards. Hence, recreational clubs are both made from their environmental publics but also make those environmental publics over time, as their practices are developed and codified by those publics that participate and as those enjoying publics adapt their behaviour accordingly.

For example, water recreation involves multiple dangers to humans because of environmental factors, e.g. pollution by pathogens, strong currents, stormy weather, and to environments because of what humans do, e.g. disturbing birds while nesting, erosion, carbon emissions through travelling to recreational sites, littering. As a consequence, recreational clubs and associations have developed codes of conduct for their members. Surfers Against Sewage (2016; Ward 1996) has a well-developed set of campaigns that encourage members to modify their own behaviour, e.g. don't drop litter, and to help the collective lobby for environmental improvements in water quality and sewage treatment. Similarly, The Green Blue (2010) is an initiative set up by the British Marine Federation and the Royal Yachting Association in 2005 to raise environmental awareness and promote sustainability by providing advice for boat users about how to behave well when boating, e.g. don't litter, minimise pollution when disposing of sewage from private boats, don't disturb wildlife.

In England, fishing clubs control the fishing rights for many lakes and stretches of river, and they enforce these through bailiffs who patrol their club's waters, checking that anglers have a valid (national) rod licence and a (local) club subscription or ticket for that site, and that anglers obey (local) site rules, such as using barbless hooks or not fishing at night. Bailiffs thus enforce environmental practices within a small group of recreational users, a role that is explicitly acknowledged, for example, when the Barbel Society are known "as the barbel police" (Eden and Bear 2012). And ordinary anglers also keep an eye on other anglers, to ensure that they follow club rules and behave properly, perhaps scolding others for bad behaviour like littering. Non-anglers may also be managed, being excluded from riparian zones by notices and bailiffing, and admonished for speeding through fords or littering.

Because of such policing, some have argued that those who spend time having fun in particular environments are more likely to act to protect them, as part of a duty of care to 'give back' to the environment that has given them so much pleasure (e.g. Taylor 2007; Sanford 2007; Snyder 2007; Tarrant and Green 1999). For example, it is common to hear angling campaigners refer to anglers as 'the guardians' or 'the eyes and ears' of the water environment and they have a history of reporting water pollution incidents.

Even recreations that have no formal organisations or clubs to police behaviour, such as walking with dogs, can have codes of conduct for environmental behaviour. The Countryside Code for land-based recreation (Merriman 2005; Parker 2006) sets out rules for behaving well in the English countryside and is especially geared at visitors from town who are presumed to be ignorant of these rules, and risk being seen as 'anti-citizens' for their unpleasant, rude or environmentally damaging behaviour (e.g. Matless 1997). Environmental codes of conduct commonly refer to practices of self-management ('keep to the footpath'), object-management ('shut the gates') and other-management ('keep your dog on a lead', 'don't feed the wild animals'), and may be communicated through physical signs on footpaths and gateways. During the 2001 Foot and Mouth Disease outbreak in the UK, more specialised rules were added locally: tubs of disinfectant appeared by gates and stiles with signs asking visitors to wash possible infection off their boots as they moved through the landscape to stop the disease spreading. Practices thus shift over time, as well as from place to place.

Similarly, anglers construct the 'anti-angler' as one who does not care about the water environment, who drops litter, who does not have a licence or pay their fees or who treats fish badly and therefore needs to be controlled by the recreational club through verbal reprimands from individuals or even full committee meetings. Garden owners in many American suburbs who fail to control their lawn in the right way, failing to mow and spray it to comply with their neighbours' expectations of tidiness, can be caught by laws stipulating weed control and other ways of correcting their perceived "civil neglect and moral weakness" (Robbins 2007, p. 99).

So 'correct' environmental practices are enforced through social norms and collective, voluntary policing by recreational clubs, shaping environmental publics who comply with (or sometimes resist) such norms. As Foucauldian 'disciplinary' measures, such exhortations and expectations "produce a public who consider their own rights as involving quiet, orderly forms of behaviour which intrude minimally with the current workings and patterns of ownership of the countryside" (Macnaghten and Urry 1998, p. 188), shaping recreationists into subjects of environmental governmentality.

Sometimes, environmental publics are recruited by the state to encourage and even enforce 'citizenly' environmental practices by others. For example, recreational freshwater anglers have long seen their club bailiffs as supplementing or even subsidising the national regulator, the Environment Agency of England and Wales, in enforcing the legal requirement for all recreational anglers to buy a valid rod licence by checking the licences for those fishing on a club's waters, as well as checking that they were club members. Budget cuts since 2008 made the support of club bailiffs even more valuable and a 'volunteer bailiffs' scheme was later formalised with the Angling Trust (2015), the national body for recreational fishing.

This also reminds us that recreational publics are diverse: at the centre of clubs are more regularly active or involved participants who draw up and sometimes police the codes of conduct, but clubs also have more occasional or lightly affiliated members who may or may not be aware of and follow those codes, and many recreational anglers may not be members of any club. Not all environmental publics subscribe to the same environmental norms or practices.

How publics perform different environmental recreations, then, is partly shaped by collective endeavours, especially when people join a club or refer to its advice. Sharing practices involves repetition but also adaptation, as outdoor recreation expands and diversifies. Despite worries about the increasingly sedentary character of daily life in urbanised societies, estimates suggest that the number of people regularly taking exercise outdoors is steady or growing, with new ways to play outdoors being invented (BASE jumping) or adapted to be more attractive (bouldering as a variant of rock climbing). And as these recreational publics grow, often their collective practices become more codified, reflected in official rules or club codes of conduct. For example, competitive outdoor swimming was first added to the list of Olympic sports for the 2008 Beijing Olympic Games, reflecting rising interest in swimming outdoors, and in 2016 marathon swimming for 10 km will be part of the Rio Olympic Games, monitored by teams of human observers and by a microchip in a bracelet that all competitors must wear to prevent cheating (Rio 2016).

Moralising practices of environmental recreation

What are the wider consequences for environmental publics of recreational engagement? Some commentators have argued that participating in environmental recreation makes publics more environmentally aware and more likely to support environmental policies, that is, that practices in different settings influence each other. For example, recreationists who engage more frequently, at length and locally with their environments are often imagined to be more responsible citizens, more knowledgeable, more caring and less hedonistic than tourists who engage infrequently and far from home with unfamiliar landscapes about which they care little after they leave: recreationists are assumed to become more environmentally enlightened citizens by virtue of their commitment to and experiences of out-door leisure. The logic here is that environmental recreationists are more aware of the environment and therefore more aware of their own impact upon it, such as through creating waste or carbon emissions, and care more about the potential damage done to environments that they enjoy, and so will behave better in envi-ronmental terms than non-recreationists.

This argument has two problems. First, the chicken and egg problem: what comes first – does someone take up outdoor recreation because they care about the envi-ronment or do they begin to care about the environment because they are already involved in outdoor recreation? Second, evidence that people link (consciously or subconsciously) their practices in different contexts in this way is patchy, although assumptions are rife, such as in Natural England's (2012, p. 58 and Table 7–3) claim that links recreational practices with consumption practices in the home:

> The more frequently people visit the natural environment, the more likely they are to appreciate it and to be concerned about environmental damage. Frequent visitors are also more likely to engage in pro-environmental behav-iours such as recycling and preferring to buy seasonal and locally grown food.

Similarly, local residents who regularly walk across a small urban greenspace may be seen positively as morally worthy guardians of that environment and more likely to care about changes to it (Waitt et al. 2009, p. 57). And people who volunteer to help on environmental projects, perhaps getting involved bodily in remaking environments through hauling stone, building walls, cutting hedges and dredging streams, especially in the countryside, are often seen as far more morally worthy and giving back to the environment that they have themselves enjoyed. But only 7 per cent of the UK adult population are estimated to get involved in environ-mental volunteering (Natural England 2012, p. 51), a much lower figure than the percentage estimated that enjoys that environment.

As well as assumptions about being greener, practices of environmental recrea-tion carry other to all sorts of moral loadings, which commentators often transfer to the publics participating without thinking about the implications. These moral loadings trade on more diffuse normative beliefs about the worth of physical exer-cise, the effort that is invested, rather than the easy gain and the greater engagement with the environment through all of the senses.

Consider the difference between walking along a country path and driving to the seaside in a private car for a picnic on the beach (e.g. Macnaghten and

Urry 1998). Walking is commonly seen as more valuable and worthy than driving, partly because enjoying a landscape by seeing it at a distance and/or mediated through a windscreen is seen as less engaged than the full-body encounter through exertion, immersion in (possibly foul) weather and even sounds and smells of the country around the walker. And there is also the ironically higher environmental impact of driving (carbon emissions, air pollution, road surfacing) to factor into this moral ordering. Speed of movement is also morally loaded: driving a car through the countryside is often denigrated as a less rich and thus less worthy encounter with that environment, because it restricts sensory connection to sight alone, but also because it produces environmental publics who are only shallowly engaged through fleeting, one-dimensional encounters, that are seen as less worthy than slower, more multiply sensed environmental practices.

Embodied practices of environmental recreation that are regularly undertaken are more readily linked to environmental learning (see Chapter 2), where enjoyment includes actively encountering, learning about and appreciating environments. For example, hikers may not only see wildlife on a walk, but hear it as well and thus learn (albeit perhaps unconsciously and patchily) about different sorts of birdsong. Indeed, the experience of sound in the countryside can itself produce different publics. For example, people who are hiking may have less tolerance of other publics going quad-biking in the same area. Restricting the speed of motor boats on Lake Windermere in England has been highly contentious precisely because environmental publics using motorised transport (motorbikes, quad bikes, motor boats) have been targeted by environmental protection measures whilst non-motorised recreations in the same spaces are not restricted. In this way, the choice of practice may well reinforce and define oppositions to other forms of practice through embodied encounters.

We can contrast the moral ordering of environmental engagement through regular recreation practices with that through the occasional encounters of tourism. Although tourism, like other forms of movement through landscapes, is embodied, often it is regarded more like consumption than recreation, because tourism prioritises seeing nice views and often paying for access, both of which are used to denigrate tourism as less engaged with the environment than more regular, repeated and prolonged forms of recreation in particular environments. For example, Urry (1995) refers to the "tourist gaze" as visual consumption of the environment in the form of scenery, thus calling up the negative associations of consumption as a shallow, self-interested and materialistic form of engagement (see Chapter 4).

Because of this, Macnaghten and Urry (1998, p. 122) argued that someone who merely consumes the environment, through tourism in particular, is denigrated as a "mere sightseer. … superficial in their appreciation of environments, peoples and places". Visual consumption is seen as dematerialising and disembodying the environmental encounter in highly negative and narrow ways. And tourist encounters with the environment, whether on land or water (e.g. Cloke and Perkins 1998; Waitt and Cook 2007) are often portrayed as fleeting, detached, shallow and insubstantial in terms of the knowledge gained of the place visited, its ecologies and cultures.

The familiar slogan of eco-tourism – take nothing but photographs, leave nothing but footprints, kill nothing but time – underlines this separation, this lack of engagement, through a lightness of touch that can be read as superficiality as well as damage avoidance.

In the USA, however, "consumptive" is used in a very different way, which is to differentiate "consumptive" activities like foraging, fishing and hunting that cause the *deliberate* extraction or destruction of environmental resources like plants and animals, from "non-consumptive" recreational activities like hiking and photography that may only cause incidental damage. However, this distinction creates some odd groupings: for example, the Department of Natural Resources (2012) in Michigan, USA, defines the following as "non-consumptive", despite their many differences: "off-road vehicle (ORV) use, snowmobiling, non-fishing related boating (including use of personal watercraft), canoeing, tubing, horseback riding, mountain biking, hiking, cross-country skiing, rock climbing, camping, mine and cave exploration, wildlife viewing, and photography". This distinction neglects the broader geographies of impact: snowmobiling uses fossil fuels so is "consumptive" of environmental resources, but those resources are produced far away from where the snowmobiling is enjoyed and are therefore not included in the Department's classification.

What is interesting here is how these environmental recreations are explicitly and institutionally separated on the basis of their consumptive purpose, despite often sharing similar footprints with notionally non-consumptive practices, especially with regard to transport. In terms of understanding practices, it is more important to see how people do things and where, especially because a single trip (for a day, a week or a month) into the environment may mingle together multiple practices, both notionally "consumptive" and "non-consumptive" in type. Someone may go fishing (a notionally "consumptive" practice), sit by a river for five hours hearing birdsong and enjoy the peacefulness but catch nothing: the fish that are notionally their object are but part of the wider environment with which they engage.

In this way, it is not the outcome, but the experience of outdoor enjoyment that makes environmental publics. Shaping environmental publics depends less on how we define the objective of recreational practices and more on how those practices are carried out and given meaning.

Place

So far, we have considered a diversity of environmental recreations and how they are differentially moralised. Let us now consider how they are *spatialized*, something that has been largely ignored by research into social practice and leisure, perhaps because leisure is seen as less serious a subject for analysis than knowledge (Chapter 2) or democratic participation (Chapter 4) or as less important in terms of the environmental impact of practices compared to those involved in consumption (Chapter 3) or work (Chapter 8). Yet people's enjoyment of the environment through recreation can be very influential in shaping environmental publics.

One way in which recreation is spatialized is in terms of its geographical location, e.g. whether it takes place at home or further away. Let us take home first. Domestic gardens are increasingly recognised as spaces for environmental recreation, sometimes conceived of as the 'outdoor room' of a house. As I mentioned above, 67 per cent of adults surveyed by Natural England (2014, p. 29) said that they regularly spent time "sitting or relaxing in my garden", although fewer (53 per cent) did some more active gardening there. So some people enjoy their garden *without* gardening, and enjoy many other recreational practices that take place there, including barbecuing, camping and feeding wildlife.

And as well as consuming food, gardens are of course important for producing food, whether individually through something as singular and private as a fruit tree, or collectively and on a larger scale, whether through community gardens, allotments and livestock rearing. In the global South, urban agriculture may make an important contribution to the diets of local low-income people who rear chickens for eggs and meat in their yards or on other unused land, whereas in the global North, it is more likely to make only a small contribution to diets, while being important in terms of pleasure and satisfaction (e.g. Bhatti et al. 2009) and socialising, volunteering and enhancing community cohesion in the case of community gardens and allotment societies. And gardening as a hobby is also big business in the global North, reminding us that environmental pleasure and consumption also support for-profit business and production (of which more in Chapter 8).

Also, environmental practices of knowing and learning are often carried out in one's own outdoor space, however small. From counting birds for national surveys like the RSPB's Big Garden Birdwatch to recording the weather, like the amateur meteorologists of the Climatological Observers Link in the UK, who submit their data to collectively produce a national record of climate (Endfield and Morris 2012), people do more than consume environments for pleasure. Environmental publics also produce data about their surroundings, which is used for various monitoring processes, feeding into wider campaigning and planning strategies, as well as into an individual's own awareness and management of those surroundings, e.g. putting out more bird food when fewer birds are seen. People who regularly (and enjoyably) take part in outdoor recreation may also be more likely to develop 'lay expertise' about environmental change (Epstein 1995), to contribute 'lay knowledge' to environmental monitoring by the state (Bell et al. 2008; Ellis and Waterton 2004) and to get involved in environmental policy debates (see Chapter 2).

What is interesting here in terms of the geographies of practice is how such records can be scaled up across multiple micro-environments into national records, producing what we might think of as a national garden biome – or more correctly a set of biomes, given the diversity of gardening practices. In a garden, publics can be made environmental, but even more obviously those publics make those environments: gardeners choose plant species, they apply fertilisers, they encourage (or discourage) animals to visit through feeding, using pesticides and landscaping, such as building ponds or walls. One might argue that these are tiny spaces, inconsequential in the larger picture of environmental change, but many small spaces add up and cause changes not only in overall land use but also in how publics are made. In his study of how lawns are grown, understood and controlled, Robbins (2007) shows how the seemingly small and mundane choices made by millions of householders in the American suburbs constitute a monocultural landscape sprayed by 23 per cent of the 2,4-D herbicide used nationally and more fertilizer than is used to grow wheat. This suburban landscape is made possible only through practices of 'lawn care' that become normalised within many neighbourhoods as what is expected of good citizens and good neighbours.

Robbins draws on Foucault, Althusser and Latour to theorise these processes, but for this book, I want to emphasise that it is what people do that makes the lawns as landscapes, but these practices also make environmental publics in the shape of gardeners who buy lawn care products, analyse the risks, operate equipment to apply the products and enjoy (or not) the results. Gardeners are therefore also learning about the environment and the risks of controlling it, as well as exposing their own bodies to the physical labour and potential pollutants involved. Making a lawn, making a garden, is thus a complex set of shared, learned and repeated practices that make publics and also make environments: the socioecological connections are both obvious yet readily neglected as merely domestic consumption. By linking knowing and doing as knowledge-practices of gardening and other outdoor recreations, we can better understand the performativity of those practices in creating new environmental realities (e.g. Law 2008; Law and Mol 2002; Waterton 2003).

Moreover, if we zoom out from the tiny lawn as one would on Google Earth, we can see the bigger picture: these multiple micro-environments are linked together through human and nonhuman networks, through hybrid flows, movements and encounters that link not only space but also time through effects on future choices. Birds visit gardens, people submit birdwatch data to the RSPB, the RSPB lobby for bird protection, people choose to put out more birdseed or nestboxes, more (or fewer!) birds visit gardens, and so on. These are flows of bodies and materials, but also digital flows of information linking spaces together through shared knowledge-practices of, about and for the environment.

As we stretch these geographies further, we see that domestic, homely spaces link also with other spaces of outdoor fun away from home. In the UK, Natural England (2012, see Table 5.4) estimated that 42 per cent of adults in England visited 'the natural environment' weekly, with most visits to the countryside (52 per cent), 38 per cent to green spaces in urban areas, e.g. parks, and 10 per cent to the seaside; 68 per cent of visits were no more than two miles away from home and 41 per cent were within one mile. Environmental recreation is therefore practised in many different places, so we also need to consider how recreation in the environment is a way of (re)making the environment, as well as how environmental publics perform their environmental playgrounds through their practices.

A frequent example used by researchers in the social sciences is walking in the countryside, perhaps because they enjoy it themselves, as well as walking being the most common environmental recreation that people report, as I discussed earlier.

TABLE 5.4 Places most frequently visited during trips away from home, ranked

Rank	Place type	Estimated visits 2011–12
1.	Park in a town/city (or other open space in a town/city)	628.4 million (221.6 million)
2	Path/cycleway/bridleway	430.1 million
3.	Woodland/forest	358.3 million
4.	Another space in the countryside	328.2 million
5.	River/lake/canal	261.4 million
6.	Farmland	241.2 million
7.	Playing field/other recreation area	228.9 million
8.	Country park	196.6 million
9.	Village	194.4 million
10.	Beach	151.8 million
11.	Other coastline	90.0 million
12.	Children's playground	80.2 million
13.	Mountain/hill/moorland	76.3 million
14.	Allotment/community garden	20.6 million

Source: Defra (2012).

Walking has been used to show how the practice produces people, through the shape of the foot (Ingold 2004), the body's sores (Wylie 2005), the interaction of boot and foot (Michael 2000) and the self-surveillance of behaviour (Matless 1997), but how it also produces environments, through deciding where to walk from/to/through, what to look at/observe while there, where to site benches, car parks, litter bins, signposts and so on, as well as developing policies to protect both paths and views for future walkers.

One environmental reality that was changed by walking as a practice that both expresses and changes environmental engagement is in the Forest of Dean Sculpture Trail in South Wales. Here, a sculpture called 'Place', which looks rather like a giant chair, draws a visitor's eye from the moment they park their car below it. When the trail was originally designed, it began at another sculpture and later wound its way up to 'Place' later, but so many people ignored the official design and instead walked first up to 'Place' because it was so very visible from the moment that they arrived in the car park, that the official trail was changed to start with 'Place', re-made by this physical movement of bodies across the landscape, pulled by vision but tracked by feet.

In similar ways, people who regularly walk familiar paths near their homes continually make and re-make the everyday relationality between people and place, between socialities and ecologies (Waitt et al. 2009). Seeing walking in the countryside as one of many spatial practices that produce different countryside spaces (e.g. Macnaghten and Urry 1998) emphasises the performativity of those practices as part of creating new environmental realities (Law 2008; also Law and Mol 2002; Waterton 2003).

But places are made not only through location but through medium. Returning to the examples with which I opened this chapter, we can consider the place of water as tied to and made through very particular and specialised environmental practices. Water offers a distinctive and attractive but problematic environment for outdoor recreation and enjoyment. Human beings live most easily on land breathing air, so moving into the water makes travelling through, seeing in and understanding the environment far more difficult for us. As well as the risk of drowning or catching debilitating diseases from contaminated water, water environments also challenge humans to make sense of and respond to invisible and unpredictable water movements.

How does one learn to live and to move through an alien environment such as deep, cold water? Through practice. Swimming, surfing, angling, boating – such outdoor recreations may use the same places but for very different practices, ranging from partial contact with water on the banks of rivers and lakes to full-body immersion, and from sedentary enjoyment on a motor-powered vehicle to aerobically intensive activity that moves an individual body slowly and laboriously through the water. This diversity of environmental encounters is what prompts Macnaghten and Urry (1998) to refer to the "multiple natures" that are produced through human-environment encounters.

The recent trend for outdoor swimming in lakes and rivers in the UK has also generated many guides to good places for indulging in this recreation. The online "Wild Swim Map" (The Outdoor Swimming Society 2016) constructs "a world of beautiful outdoor swimming spots" from rivers and lakes that would either be invisible to a passer-by or be seen as places for fish, boulders, mud and storm runoff, rather than places for people. Here, the practices of swimming change both the meaning of place and, if more people come to use it as a consequence, also change the materiality of place too, producing new geographies of possible human–water encounters.

Other ways of spatializing environmental recreation are less embodied and more vicarious, re-imagining environments as places for recreation through online and offline media. Guidebooks and websites offer photographs, videos, advice, maps and narratives about where to find and how to choreograph practices of outdoor recreation, sharing these geographies through writing and reading, watching and listening, recording and posting. These geographies are also often differentiated by time and space, specific to particular cultures or time periods. As mentioned at the beginning of this chapter, the English Lake District was seen by Daniel Defoe as horrible and frightening in the 18th century, but by the early 19th century, it had become fashionable to value such landscapes. The poet, William Wordsworth, first published his *Guide to the Lakes* in 1802, urging visitors to appreciate the region not through scenic set pieces and established artistic terminology, but through the unity and sublimity produced by nature itself – something that needed to be experienced (and was experienced in his case) through walking, emphasising the embodied encounter with the landscape as the environment that surrounds and permeates us, rather than as landscape that is framed and mounted on a wall.

These representational practices are particularly important today on a global level, as outdoor recreation expands, diversifies and spreads globally through tourism and travel. Countries far away from the home of the recreationists can be identified, mapped and evaluated as environmental playgrounds, with 'virtual' trips possible from one's armchair through watching interactive videos online, posted by official organisations promoting tourism or by individual visitors. Imagining faraway places is facilitated by online materials, as indeed it has been through paper materials and word of mouth for centuries, although today outdoor recreations such as bungee jumping, swimming with dolphins and whale watching take very different approaches to natural spectacles than was typical of the past. Geographies of ecotourism and environmental experience continually invent new modes of recreation, such as snowboarding or sand-yachting, that take place on the same mountains or beaches as older modes, such as hiking and swimming.

As well as imagining and perceiving, recreational practices spatialize in other ways. Codes of conduct, as discussed earlier, vary by practice and by geography: some clubs are national, but others are local or regional and have their own rules and customs and may police these codes through reporting, bailiffing and prosecution as ecodisciplinary measures, thus shaping what people do in environments, particularly locally. And recreationists also make places through embodied labour: anglers build jetties, dredge rivers and stock lakes with juvenile fish; gardeners sow, dig, compost and build on their plots; bird enthusiasts create and extend reed beds, shaping hydrologies and ecologies. Re-landscaping places for outdoor recreation thus changes how environments look and function, as well as linking the material world of soil, rock, plants and water with imaginaries of future environments.

Power

Let us now consider how powerful these diverse recreational publics are. This is an area of strong contradictions. First, many recreations are dismissed as unimportant because they are merely hobbies, and thus not worthy of serious consideration, or because they are minority interests, with only small percentages of the population involved. Certainly, as we saw earlier in this chapter, some outdoor recreations only involve small minorities, but often still number in the hundreds of thousands, and adding together different recreations that use the same space, e.g. walkers, cyclists, horse-riders all using the same footpath across a moorland, thereby producing much larger numbers.

Second, recreation is often seen as merely a form of consumption, so that recreationists, like consumers, are regarded as powerless to change or protect environments, being unknowing and unfeeling about the nature that they encounter. Tourism is a good example of how recreation and pleasure is rendered as consumption, especially when efforts are made to monetarise the importance of outdoor recreation in terms of how many millions of pounds a national park or other environmental amenity contributes to the local or national economy by prompting tourists to visit it and spend money there while doing so. Here, environmental engagement is expressed not in terms of emotional connection or specialised knowledge but in terms of monetary exchange – how much publics 'spend' in the environment becomes a symbol of their commitment and of the importance to the regional or national economy of the environment as a playground.

Monetarising multiple tourism visits is difficult in practice, and there are whole literatures in the discipline of economics and related research fields that explore methods to estimate the worth of particular environmental assets – a national park, elephants, sea turtles, a rainforest – to local tourism. Such exercises arose because of convictions that natural resources could be better protected if a stronger case were made for them in monetary terms, because their losses could then be set against the monetary gains from new developments that might impact them, e.g. new airports or roads. As a corollary, monetary values can be invoked by government departments with environmental responsibilities to legitimate environmental protection, especially where this competes with other responsibilities or the plans of other departments. 'Ecosystem services' (e.g. UK National Ecosystem Assessment 2011) is the latest incarnation of this way of valuing environments in order to gain political purchase against other, more clearly economically productive policy areas.

This argument assumes – fairly reasonably, some might say – that framing decisions in monetary terms is so common in modern decision-making that proposals that cannot be framed in monetary terms will inevitably lose out to ones that can. The language of consumption, of buying one's way, of commoditising natural enjoyment, is both pervasive and pejorative, despite many engagements with environments not being motivated by money, but nevertheless being highly important for many people. The problem is that the figure of 'the consumer' is seen by many as powerless politically to challenge the juggernaut of modern capitalism in exploiting environmental resources. By comparison, the figure of 'the citizen' is seen far more positively as politically engaged, potentially powerful and motivated by stronger and more worthy principles than monetary ones.

Third, and underlying these arguments, is the dualism of citizen versus consumer that we have met before, the assumption that two mutually exclusive modes of engaging in the modern world are defined in opposition and hierarchically. This assumption renders 'the consumer' less worthy than 'the citizen', because the citizen is part of the social collective and acts in support of democracy and the wider public interest, whereas the consumer behaves in a shallow and selfish way, as a sort of 'anti-citizen'.

Practices are also differently moralised in terms of how they are performed, with embodied recreational encounters often given stronger moral worth. Physical presence can also contribute to influence through political protest, e.g. in the 'mass trespasses' of the 1930s in the UK when hundreds of people walked across private moorlands in the central uplands of England, as part of a wider campaign for public access to the environment. The Ramblers Association (now 'The Ramblers') was one of the organisations established to campaign for access rights to the environment for recreation, developing a politics of amenity that carried on through the 20th century until the Countryside and Rights of Way Act (the CROW or 'right to roam' legislation) was enacted in the early 21st century and on through the process of mapping out the new 'access' areas. In terms of power, amenity protests benefited from the embodied, personal resistance offered against the institutionalised control and exclusiveness of private lands, and the presence of bodies in the landscape also gave visual evidence of the strength of public opinion.

Despite these changes in public access to land, other types of access remain highly contentious. Access to rivers in England is hotly contested between anglers, canoeists and (motor) boaters, making watery places battlegrounds for recreational disputes. Because of this, recreationists are often co-opted into policy communities, to sit on committees charged with deciding how best to manage valued environments (see Chapter 2), and thus may have some direct influence over policy-making, although this is often restricted to the local scale.

A quite different powerplay is involved where specialised publics manage particular environments themselves. Rather than protesting to gain access to private lands, some recreational clubs buy or lease the recreational rights to environments and manage them in their own interests. For example, the UK's Royal Society for the Protection of Birds (RSPB) buys and manages nature reserves to increase bird populations so that members can enjoy watching them. Allotment holders grow fruit and vegetables in their own plots and community or 'guerrilla' gardeners (unofficially) adopt urban wastelands to plant attractive foliage and small-scale food crops. Angling clubs lease the riparian rights on particular lakes and river banks and then manage these areas to increase populations of target fish, by physically re-landscaping environments – river banks are re-contoured or reinforced to correct perceived problems of banks eroding or to diversify water flow, weirs and groynes are built of wood, stone or concrete, to change and diversify flow; gravel may be added to improve fish spawning. They may remove vegetation, dredging out weed or woody debris, or pulling out plant species identified as 'invasive exotics', or add vegetation, especially plant species identified as 'native', e.g. anglers have planted willow, ash, oak, alder, hawthorn trees along the River Swale in northern England, changing the riparian landscape from grassy levees to wooded banks (see Figure 5.1). All of this uses various assemblages of the bodily labour of anglers, their boats and other tools such as rakes, chainsaws and herbicides, as well as powerful earthmoving equipment such as bulldozers. As well as managing water, rocks and plants, angling clubs manage animals, shooting or scaring away predators such as cormorants, goosanders and mink, blamed for reducing fish populations and thus catches, but pouring thousands of juvenile fish into lakes and rivers where they perceive target populations are at risk.

In these ways, recreationists can be seen as 'performative publics' (Eden and Bear 2012), who literally reshape environments, ecologies and hydrologies for recreational access and pleasure through 'hands-on' lay management of environments that they value and enjoy, (re)making environmental realities as well as themselves.

Of course, not all recreationists get involved in all of these activities. An angling club of several hundred people may include only a few dozen who actively and physically participate in management efforts and members may vary greatly in how often they visit the waterways managed by their group and how long they spend

FIGURE 5.1 Willows planted by anglers on the River Swale in northern England.

there each time they do visit. What is important for this book is that recreational environmental publics do not merely consume the environment in a one-way, sometimes passive interaction; instead, they often also produce the environment in a far more active, even creative relationship as part of a wide assemblage of humans, animals, plants, devices, water and earth.

As I discussed earlier in this chapter, recreational organisations and clubs can shape their communities' behaviours by producing and circulating environmental information about where and how to enjoy the environment safely, about pollution and environmental change and about campaigns to protect environments, developing codes of conduct for their members as collective norms for environmental behaviour, risk assessment and training. Environmental codes of conduct produced by clubs and shared habits of environmental encounter help to produce 'specialised publics', even as they address and seek to represent them (Warner 2002).

And where recreationists have felt let down by state agencies and their environmental policies, they have set up campaigns or independent organisations, such as the Anglers Conservation Association (Bate 2001; now Fish Legal) and Surfers Against Sewage (Ward 1996) to lobby the state to better protect environments. Such activities have prompted very little research attention from social scientists, despite their considerable potential to give insight into how people think about and come to care for particular environments. This again emphasises that recreational publics are not commonly regarded as having power and influence, despite their practices continually giving meaning to and remaking everyday environmental realities.

Summing up

This chapter has emphasised how environmental publics are made through outdoor recreational practices, places and relationships. These practices are widely divergent in terms of time invested, location, physicality, bodily immersion and numbers participating, but the personal, embodied engagement with the environment that they offer contrasts strongly with practices analysed in other chapters, such as participating (Chapter 3) and voting (Chapter 7). Their geographies of practice spread from the local, familiar environment of the private garden through public access to open land and to international tourism and the virtual spaces of the internet, bringing different worlds of environmental leisure to the armchair surfer.

Unfortunately, much of this has been neglected by research into outdoor leisure, which instead has focussed upon measuring satisfaction and enjoyment and the benefits to humans in terms of physical and mental wellbeing, as well as bodily shaping and gendering. Little of this literature considers how recreational practices are also (or might be) environmental, although this is beginning to be more considered, or how they are spatialized and linked to hands-on management. There is, however, a lot of moralising about recreations that sometimes obscures the actuality of practice. When recreationists are cast merely as consumers of environmental scenery and other amenities, rather than powerful agents of environmental remaking through their intimate, regular encounters with environments, this denigrates recreational practices to the realm of sensual rather than constitutive reality.

To counter this, I have emphasised how environmental recreations are learned, shared, developed, enforced, resisted and codified through collective endeavours and through diverse, often practice-specific assemblages of publics, environments, equipment and meanings. Through becoming recreationists, publics are themselves enacted environmentally but also contribute to a wider understanding and management regime for those environments which they seek to enjoy and thus also seek to protect and guard.

Sometimes, the experience of enjoying active environmental recreation prompts recreationists to become active campaigners for environmental reform, as their enjoyment becomes more politicised. This leads us nicely into the more detailed discussion of such practices in the next chapter.

Note

1 I exclude holiday recreations because they are uncommon (occasional) and less important in terms of making environmental publics through regular, repeated, routinized practices.

References

Angling Trust (2015). Voluntary bailiff service. www.anglingtrust.net/page.asp?section=93 0&sectionTitle=Voluntary%20Bailiff%20Service

Bate, R. (2001). *Saving Our Streams: the role of the Anglers' Conservation Association in protecting English and Welsh rivers.* Institute of Economic Affairs, London.

Barratt, Paul (2011). Vertical worlds: technology, hybridity and the climbing body. *Social & Cultural Geography* 12, 4, 397–412.

BBC (2010). Cumbria Great North Swim cancelled over safety fears. www.bbc.co.uk/news/uk-england-cumbria-11175208

Bell, Sandra, Mariella Marzano, Joanna Cent, Hanna Kobierska, Dan Podjed, Deivida Vandzinskaite, Hugo Reinert, Ausrine Armaitiene, Malgorzata Grodzińska-Jurczak and Rajko Muršič (2008). What counts? Volunteers and their organisations in the recording and monitoring of biodiversity. *Biodiversity Conservation* 17, 3443–3454.

Bhatti, Mark, Andrew Church, Amanda Claremont and Paul Stenner (2000). "I love being in the garden": enchanting encounters in everyday life. *Social & Cultural Geography* 10, 1, 61–76.

Blomley, Nicholas (2004). Un-real estate: proprietary space and public gardening. *Antipode* 36, 4, 614–641.

Cloke, Paul and Harvey C. Perkins (1998). "Cracking the canyon with the awesome foursome": representations of adventure tourism in New Zealand. *Environment and Planning D: Society and Space* 16, 185–218.

Defra (Department for the Environment, Food and Rural Affairs) (2011). www.defra.gov.uk/statistics/files/Statistical-Release-13-April-2011-biodiversity1.pdf (now deleted).

Defoe, Daniel (1724–1726). A tour thro' the whole island of Great Britain, Letter 10: Lancashire, Westmorland and Northumberland. http://www.visionofbritain.org.uk/travel/as.Defoe/34.

Department of Natural Resources (2012). Social & economic overview. Michigan, USA. www.michigan.gov/dnr/0,1607,7-153-10370_30909_43606-153362--,00.html#recreation (now deleted).

Eden, Sally (2009). Commentary: environmental solutions. *Environment & Planning A* 41, 2037–2040.

Eden, Sally and Christopher Bear (2012). The good, the bad, and the hands-on: constructs of public participation, anglers, and lay management of water environments. *Environment and Planning A* 44, 1200–1240.

Ellis, Rebecca and Claire Waterton (2004). Environmental citizenship in the making: the participation of volunteer naturalists in UK biological recording and biodiversity policy. *Science and Public Policy* 31, 2, 95–105.

Endfield, Georgina H. and Carol Morris (2012). Exploring the role of the amateur in the production and circulation of meteorological knowledge. *Climatic Change* 113, 1, 69–89.

Epstein, Steven (1995). The construction of lay expertise: AIDS activism and the forging of credibility in the reform of clinical trials. *Science, Technology, & Human Values* 20, 4, 408–437.

Great Swim (2011). About the Great North Swim. www.greatswim.org/Events/British-Gas-Great-North-Swim/AboutTheGreatNorthSwim.aspx

The Green Blue (2010). The green blue. http://thegreenblue.org.uk/

The Guardian (2011, 13th September). David Walliams swim in river Thames makes a million for charity. www.guardian.co.uk/tv-and-radio/2011/sep/12/david-walliams-swim-million-charity?intcmp=239

Hartig, T. and Marcus C. Cooper (2006). Healing gardens: places for nature in health care. *The Lancet* 268, 536–537.

Hartig T., G. W. Evans, L. D. Jamner, D. S. Davis and T. Gärling (2003). Tracking restoration in natural and urban field settings. *Journal of Environmental Psychology* 23, 109–123.

Hinchliffe, Steve, Matthew B. Kearnes, Monica Degen and Sarah Whatmore (2007). Ecologies and economies of action: sustainability, calculations, and other things. *Environment and Planning A* 39, 2, 260–282.

Ingold, Tim (2004). Culture on the ground: the world perceived through the feet. *Journal of Material Culture* 9, 3, 315–340.

Law, John (2008). On sociology and STS. *The Sociological Review* 56, 4, 623–649.

Law, John and Annemarie Mol (2002). Complexities: an introduction. 1–22 in John Law and Annemarie Mol (edited), *Complexities: social studies of knowledge practices*. Duke University Press, Durham.

Macnaghten, Phil and John Urry (1998). *Contested Natures*. SAGE, London.

Matless, D. (1997). Moral geographies of English landscape. *Landscape Research* 22, 141–155.

Merriman, Peter (2005). "Respect the life of the countryside": the Country Code, government and the conduct of visitors to the countryside in post-war England and Wales. *Transactions of the Institute of British Geographers* 30, 3, 336–350.

Michael, Mike (2000). These boots are made for walking…: mundane technology, the body and human-environment relations. *Body & Society* 6, 107–126.

Natural England (2012). Monitor of engagement with the natural environment: the national survey on people and the natural environment. Annual report from the 2011–12 survey. www.naturalengland.org.uk/ourwork/enjoying/research/monitor

Natural England (2014). Monitor of engagement with the natural environment: the national survey on people and the natural environment. Annual report from the 2013–14 survey. www.gov.uk/government/statistics/monitor-of-engagement-with-the-natural-environment-2012-to-2013

The Outdoor Swimming Society (2016). *Wild Swim Map*. http://wildswim.com

Parker, G. (2006). The Country Code and the ordering of countryside citizenship. *Journal of Rural Studies* 22, 1–16.

Pretty J., J. Peacock, R. Hine, M. Sellens, N. South and M. Griffin (2007). Green exercise in the UK countryside: effects on health and psychological well-being, and implications for policy and planning. *Journal of Environmental Planning and Management* 50, 2, 211–231.

Rio 2016 (2016). Marathon swimming. www.rio2016.com/en/marathon-swimming

Robbins, Paul (2007). *Lawn People: how grasses, weeds, and chemicals make us who we are*. Temple University Press, Philadelphia, PA.

Sanford, A. Whitney (2007). Pinned on Karma Rock: whitewater kayaking as religious experience. *Journal of the American Academy of Religion* 75, 4, 875–895.

Shove, Elizabeth and Mika Pantzar (2005). Consumers, producers and practices: understanding the invention and reinvention of Nordic walking. *Journal of Consumer Culture* 5, 1, 43–64.

Snyder, Samuel (2007). New streams of religion: fly fishing as a lived, religion of nature. *Journal of the American Academy of Religion* 75, 4, 896–922.

Sport England (2009). *How We Recognise Sports*. www.sportengland.org/about_us/recognised_sports/how_we_recognise_sports.aspx (accessed 17 Dec 2009).

Surfers Against Sewage (2016). *Big Spring Beach Clean*. http://www.sas.org.uk

Tarrant, Michael A. and Gary T. Green (1999). Outdoor recreation and the predictive validity of environmental attitudes. *Leisure Sciences* 21, 17–30.

Taylor, Bron (2007). Focus introduction: aquatic nature religion. *Journal of the American Academy of Religion* 75, 4, 863–874.

Taylor, Bron (2007). Surfing into spirituality and a new, aquatic nature religion. *Journal of the American Academy of Religion* 75, 4, 923–951.

UK National Ecosystem Assessment (2011). http://uknea.unep-wcmc.org/Resources/tabid/82/Default.aspx

Urry, John (1995). *Consuming Places*. Routledge, London.

US Fish and Wildlife Service (2012). *Hunting Statistics and Economics*. www.fws.gov/hunting/huntstat.html

Waitt, Gordon and Cook, Lauren (2007). Leaving nothing but ripples on the water: performing ecotourism natures. *Social & Cultural Geography* 8, 4, 535–550.

Waitt, Gordon, Gill, Nicholas and Head, Lesley (2009). Walking practice and suburban nature-talk. *Social & Cultural Geography* 10, 1, 41–60.

Ward, Neil (1996). Surfers, sewage and the new politics of pollution. *Area* 28, 3, 331–338.

Warner, M. (2002). Publics and counterpublics. *Public Culture* 14, 49–90.

Waterton, Claire (2003). Performing the classification of nature. 111–129 in Bronislaw Szerszynski, Wallace Heim and Claire Waterton (edited), *Nature Performed: environment, culture and performance*. Blackwell, Oxford.

Wylie, John (2005). A single day's walking: narrating self and landscape on the South West Coast Path. *Transactions of the Institute of British Geographers* 30, 2, 234–247.

6

CAMPAIGNING PUBLICS

Introduction

In July 2010, I was one of about fifty people who attended a meeting of my local council. We were there to show our opposition to the council's plans to award a twenty-five-year contract to hand over management of the future municipal waste generated by residents in its own area and that of an adjacent local authority to a multinational company that intended to build a very large incinerator to burn much of this waste. Our presence was coordinated not by any large non-government organisations, although some (such as UKWIN and Friends of the Earth) had supported us with information and advice, but by our parish council and neighbours.

We stood outside the front door of the town hall with placards, to try to influence the councillors as they came in, then most of us took seats in the public gallery (where we were not allowed to speak) while a few gave three-minute speeches to try to persuade the councillors to vote against the plans. We failed and most of the councillors voted for the contract. The campaign continued for another four years, involving more demonstrations (Figure 6.1), speeches and demonstrations at council meetings, as well as email petitions, letter writing, fund-raising, commissioning reports by environmental consultants, seeking judicial review, employing a barrister to argue for permission to appeal the decision not to allow a judicial review and participating in the community liaison committee described in Chapter 3.

All of these efforts failed and construction work on the incinerator site began in 2015. A new liaison committee was convened with the intent of distributing funds to local community works, as promised in the planning application, and some campaigning in the form of letter-writing and emailing continued as the ongoing construction disrupted traffic and was blamed for noxious odours.

FIGURE 6.1 Anti-incinerator protests outside County Council offices in northern England.

This chapter is not about incinerators, planning processes, councils or even non-governmental organisations. Instead, it considers how we can understand the environmental publics that form through these campaigns, the kind of practices that they adopt, invent and share, how they are changed by enacting those practices and the implications of those campaigns for the spatialisation and power of protest. In Chapter 3, I looked at publics that participate in environmental planning and decision-making, but focussed on how publics were drawn into those processes or sought to change them, especially through public-government relationships. In this chapter, I look at how publics *become* campaigners, especially through grassroots mobilisation, artistic expression, bodily occupation and even institutionalisation. In other words, rather than thinking about how environmental publics are made through their involvement with state processes, this chapter is about environmental publics making and sharing campaigning practices, as well as the implications for geography and power. I emphasise how such organisations are made and given voice as expressions of public unease and imagining, as well as how they often struggle to sway better-resourced opponents.

First, a word about terms. Environmental groups created by the public (as opposed to groups created by government or by corporations for liaison purposes, as we encountered in Chapter 3) are variously referred to as 'non-governmental', 'community-sector' or 'third-sector' organisations – NGOs, CSOs or TSOs respectively. Some of these labels define the practices that we shall discuss in this chapter by what they are *not*, rather than what they are, as they are neither in the private sector of profit-making companies nor the public sector of state institutions. Writers often also refer to 'civil society' to encompass these various groups, as well as individual citizens who may not formally belong to any group but nevertheless share their values or campaigns. Yet all three 'sectors' (government, private, non-governmental) are internally heterogeneous, so drawing clear boundaries is often difficult, for example, where NGOs sell things through online marketing or where large private companies set up not-for-profit or charitable offshoots.

The amount of 'organisation' shown by non-governmental organisations also varies greatly, from loosely affiliated 'grassroots' groups that meet informally and rely on local friendship networks, to more formalised organisations with multi-million pound budgets, office buildings and full-time staff. 'Grassroots' groups are also often seen as embryonic, non-hierarchical and amorphous, maybe emerging in opposition to something like an incinerator or housing development, and sometimes ephemeral, fading away as the object of opposition either itself fades or is built, making further opposition pointless. In contrast, more formalised NGOs are likely to have increasingly professionalised staff and structured systems of managing people (e.g. human resources departments) and money (e.g. accountants and investment managers), as well as greater liabilities and longer lifetimes.

Despite this, any organisation can still disband, merge into another or collapse. Campaigns therefore do not only begin and grow, they also merge or die and are lost to the environmental scene. For example, in Eastern Europe in the 1980s, grassroots environmental organisations took advantage of a political opportunity: environmental politics was one subject that was not blacklisted for public discussion, unlike many others. Hence, the environmental movement in countries like Poland and Hungary protested against the fallout from Chernobyl, dams on the Danube and other environmentally damaging developments and led demands for political change. But after the collapse of the Berlin Wall and the opening up of eastern European countries to western-style democratic politics and consumer lifestyles, this role declined. Environmental movements in the Baltic states, Ukraine, Slovakia and Moldova gradually merged with nationalist parties and elsewhere they became increasingly professionalised in the 1990s, moving away from the grassroots and becoming more like the NGOs that had been established in western Europe for decades (Jancar-Webster 1998, p. 87).

Another term used about campaigning publics is 'new social movements'. The term 'social movement' emphasises mass participation by citizens in political action and ideology, especially where a range of organisations are loosely networked because they share the same values. They are called 'new' where they do not draw on the 'old' class divisions between labour and capital, as did political parties and trade unions, but on cross-cutting concerns about war, nuclear power, free speech, the environment, race, gender equality and sexuality, concerns that became politically widespread in many countries in the later 20th century.

But I will avoid referring to 'the environmental movement' as others have (e.g. Lowe and Goyder 1983; McCormick 1992; Nicholson 1989), because that term suggests too great a degree of coherence between groups over values and practices; instead, I will emphasise the diversity of campaigning practices. Another reason to avoid this term is to avoid the theoretical debate about what makes a 'new social movement' and which organisations are part of it and which are not. Instead, this chapter will focus on how environmental publics perform campaigning, how this shapes their views of themselves and other people's views and expectations, as well as what this tells us about the geographies and power relations of environmental publics.

Practices

This chapter will therefore bring together and make sense of a ragbag of shifting practices that are enacted by and shape public campaigning for the environment. I will start by considering the sorts of environmental publics that are involved in campaigning in terms of their size and practices.

First, we can contrast what we might call the 'core' and the 'periphery' of campaigning groups. The 'core' is a very small proportion of the public who are actively involved in environmental campaigning and are at the centre of many voluntary groups. These activists often begin campaigning as volunteers, but as NGOs grow and can recruit professional staff, from accountants to advertisers, volunteers may also move into paid positions where campaigning is their day job. Those who have made this move are less relevant to my focus in this book than the 'periphery' of environmental NGOs, that is, those publics not at the centre of groups but who may get involved sporadically or follow group activities from the sidelines.

In their 1983 book, Philip Lowe and Jane Goyder referred to such people as "the attentive public", meaning people who were not officially members of environmental campaigning groups but nevertheless shared their values. This is a useful term and Lowe and Goyder (1983, p. 9) also helpfully defined it through practices (although they did not use that word) such as

> readership of various environmental magazines, students of environmental studies in schools, colleges, and universities, sympathetic members of the design and land-use professions and the many people who, through their personal convictions, behaviour and life styles, express their concern for the environment – for example organic gardeners, health food devotees, outdoor enthusiasts and supporters of recycling schemes.

To borrow a metaphor from consumption, the attentive public are an important 'market' that many environmental campaigning groups try to reach, both to sell memberships and other products but also to mobilise support for other (non-monetary) practices, e.g. encouraging people to sign petitions or write letters and emails to their MPs.

Figure 6.2 turns these ideas into a pyramid to demonstrate the relative size of each group. The smallest group is those actively involved in environmental campaigning groups – the 'core' of activists at the top of the pyramid.[1] The next level down is larger, because many more people are members of environmental groups than are core activists; such members may pay formal subscriptions, be sent newsletters and

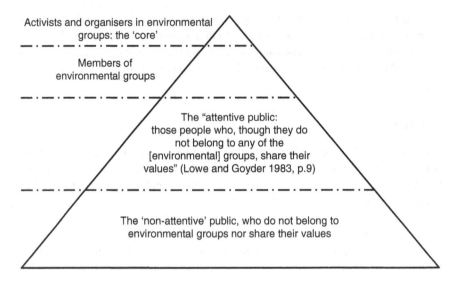

Activists and organisers in environmental groups: the 'core'

Members of environmental groups

The "attentive public: those people who, though they do not belong to any of the [environmental] groups, share their values" (Lowe and Goyder 1983, p.9)

The 'non-attentive' public, who do not belong to environmental groups nor share their values

FIGURE 6.2   Pyramid of activist, member, attentive and non-attentive publics, indicating comparative sizes.

attend demonstrations, but do not regularly participate in activism as a day job. This is especially true if we include amenity groups such as the National Trust in the UK, which has millions of members who may have joined to benefit economically from discounts on fees to visit its popular heritage and environmental sites, rather than to support the Trust's agenda for preserving and restoring valued landscapes that it owns.

However, getting accurate figures for the number of people who belong to environmental campaigning organisations is notoriously difficult because many organisations do not regularly release membership figures. Even if they do, figures from different organisations cannot simply be added up to give a national total because many people belong to more than one organisation, so instead researchers have tended to use surveys to ask people if they belong to environmental organisations.

For example, in a 2000 US poll (Dunlap and McCright 2008), 5 per cent of people surveyed reported belonging to an environmental organisation (Sierra Club and Greenpeace were named as national and international examples respectively in the survey question) and 9 per cent reported belonging to a local, regional or state environmental organisation, with some overlap where people reported belonging to both, suggesting that about 11 per cent of people belonged to some sort of environmental organisation. However, some said that they did belong to environmental organisations but also said that they were unsympathetic to the environmental movement. Dunlap and McCright (2008) speculated that this mismatch happened because these people were members of environmental sports organisations, such as hunting clubs, which approached environmental issues very differently from environmental protection NGOs. This illustrates nicely the problems with the term 'the environmental movement', because one person's idea of what the environment should be (a landscape full of game to hunt) might be very similar to another's (a landscape full of animals to watch) but they might use very different labels for their environmental ideas and practices.

In international surveys of European, Latin American and East Asian nations reported by Dalton (2005, 2015), about 5–10 per cent of people surveyed said they were members of environmental groups, with small variations by country.[2] Slightly fewer people said that they had actually participated in environmental protest. These compare well with the US results and also with the proportion of people self-reporting membership of political parties (6.5 per cent), of community groups (5.2 per cent) or of other campaigning groups such as women's (4.9 per cent) or peace groups (2.7 per cent). However, we again need to remember that these are self-reported figures, not observational figures, so will include inaccuracies where people forget they belong to an organisation or that they report belonging to an organisation although their membership has lapsed or they get confused or they do not tell the truth for whatever reason.

In any case, most people who *are* members of environmental campaigning groups are members of the larger groups, which are fewer in number and also less typical of environmental campaigning groups in general. In a survey of 139 environmental groups with over 4.5 million members and supporters in total, the largest twelve groups accounted for more than 80 per cent of the members (Cracknell et al. 2013). About a third of the income of all 139 organisations came from individual people through donations, membership fees, entry fees and the like (Cracknell et al. 2013), emphasising again how much campaigning organisations depend upon environmental publics to keep going. For example, Greenpeace UK (2014) reported income over £15 million in their 2013–14 accounts, of which over £13 million came from supporters and legacies, that is, from supportive environmental publics rather than government grants or sales of merchandise.

As the pyramid in Figure 6.2 suggests, there are far more members of the public who are members and supporters of environmental NGOs than there are professional campaigners staffing those NGOs. In total, the 139 organisations surveyed by Cracknell et al. (2013) had only 11,125 employees (full time-equivalent) working on environmental issues, against over 4.5 million ordinary members or supporters. And as well as giving money to groups for campaigning costs, having many members is helpful to groups because it suggests that their arguments are well supported by the wider population, giving them greater legitimacy and their arguments more weight when lobbying for change or seeking media coverage.

But does belonging to an environmental NGO affect other sorts of environmental practice, such as signing petitions and lobbying governments? Jordan and Maloney (1997) argue that it does not, because buying membership in an environmental group is more like a purchase than a commitment to political action with that group. Based on a survey of Amnesty and Friends of the Earth in the UK, Jordan and Maloney (1997) argued that members of such campaigning NGOs do little to participate in the campaigning work itself, either because they are not interested in political action on these issues (over 70 per cent of FoE members surveyed said that

TABLE 6.1 Membership of selected environmental NGOs in the UK, ranked by income

Name	Membership			Income 2013–2014	Environment staff in full-time equivalents 2013–2014
	1971	1991	2013–2014		
The National Trust (England, Wales and Northern Ireland)	278,000	2,152,000	4,065,579 individuals*	£435,918,000	5,285**
The Wildlife Trusts	64,000	233,000		£137,000,000	2,090
RSPB (Royal Society for the Protection of Birds)	81,000	852,000	1,114,938	£119,677,000	1,683
WWF UK	12,000	227,000	421,000 "regular givers"	£66,177,000	312
Sustrans	–		40,000 supporters	£48,872,000	183
Woodland Trust	–	63,000		£31,878,704	291
Greenpeace UK	–	400,000		£13,151,726	89
Friends of the Earth UK	1,000	240,000		£10,147,715	146
The Ramblers	22,000	87,000	107,584	£8,906,000	69**
CPRE (Campaign to Protect Rural England)	21,000	46,000	62,000	£4,172,000	36**

Source: some figures were taken from the University of Birmingham's research project on environmental NGOs, see www.ngo.bham.ac.uk/appendix/Green_Membership.htm, but others were taken from NGO websites during 2015.

Notes:

* 2,020,585 individual subscriptions.

** These figures represent all staff, not solely environmental staff. Many organisations do not release data consistently, that is, annually and in the same format, so that is why there are some gaps in this table.

being politically active was not an important reason for them to join FoE (p. 191) or because the hierarchical structure of the organisation prevented them by restricting political action to those elites on the organisation's professional staff (p. 188).

Clearly, as the pyramid in Figure 6.2 suggests, I do not disagree that most members of environmental groups are not regularly active in environmental campaigns. However, by focussing more on practices, I see involvement in campaigning as more flexible than Jordan and Maloney (1997) suggested. Over time, some people may pay their subscription to belong to a NGO for a while without taking part in any of its protest actions and then leave, becoming inactive, or they may continue to subscribe but also get more active in protest actions and, perhaps in response to a new planning proposal, set up a new group to campaign in ways that the original group did not. Hence, there are different stages along a spectrum of campaigning activity for environmental publics, a spectrum that also climbs the pyramid in Figure 6.2, from the inactive public, to the active public, to the informal and small NGO, up to the highly professionalised and large NGO. Moving between stages involves not only learning or inventing new practices, but also changing one's own identity and the meanings of what it is to be an activist. Progression up and down this ladder is neither automatic nor inevitable, and may even go into reverse as people become disillusioned with campaigning or their personal lives change so that they no longer have the time, ability or willingness to devote to a cause.

Attentive environmental publics

That suggests that the line in Figure 6.2 between 'members' and 'attentive publics' oversimplifies the distinction. For example, someone may take up a subscription to an environmental NGO and then stop it two years later without any serious decline in their general attention to environmental issues, for instance. So to include all environmental publics, we should move down the pyramid in Figure 6.2 to the 'attentive public', that is, those people who are not 'members' of nor 'activists' in environmental NGOs.

Again, the size of this 'attentive public' is difficult to estimate. Inglehart (1990) estimated that around 8 per cent of people surveyed in western countries considered themselves to be members of or potential members of what he called the "ecology movement", although there were large geographical variations by country.[3] In the USA, Dunlap and McCright (2008, p. 1049–1051) reported that around 16 per cent of those surveyed in 2000–2006 thought of themselves as "an active participant in the environmental movement" and around 50–55 per cent thought of themselves as "sympathetic towards the movement, but not active" in it, which we can interpret as a rough measure of the 'attentive' public. So while activist publics are few in number, attentive publics are probably in the majority, with some variations across time and space, for example, declining somewhat following the 2008 global recession.[4]

What kinds of publics are these? Various surveys have attempted to correlate sociodemographic characteristics with environmental sympathies. Women and those self-identifying as politically liberal in surveys were more likely to self-identify as 'active' or 'sympathetic' to environmental groups (Dunlap and McCright 2008, p. 1058). Wealthy people were also more likely to report environmental campaigning, with Dalton (2015, p. 539; 2005) finding that self-reported environmental activism was strongly correlated with a country's economic development as measured by GNP per capita and the UN's Human Development Index, and also (but slightly less strongly) correlated with indices of political democracy. Age is also implicated: Inglehart (1990) found that students tend to report more support for environmental activism than non-students and Dalton (2015) found that young people tend to report being more environmentally active than older people. Lowe and Goyder (1983) argued that the environmentally 'attentive public' is often more middle-class, although this is also true for those attending to non-environmental campaigning and thus a function not of environmental issues, but of more general trends in social capital, leisure time, education, income and location.

McNeish (2000) surveyed 'grassroots' groups who campaigned against road proposals and were members of Alarm UK, an umbrella national NGO providing support, information and advice. Although based on self-selection, his results (ibid., p. 189) suggested that the groups were predominantly white, more male than female, mainly under 45 years old and had completed far more education than the national average (68 per cent had a university degree). In terms of political thinking and practice, they reportedly were more likely to lean to the left (60 per cent), more likely to be members of a political party (35 per cent) and far more likely to be members of another environmental group (62 per cent reported belonging to Friends of the Earth, 32 per cent to Greenpeace, 20 per cent to Earth First!, for example) than the national average. Those working in 'caring' professions, such as teachers, social workers and medical professionals, also are more likely to belong to environmental groups (e.g. Lowe and Goyder 1983, p. 10).

But such estimates present several problems. First, correlations between reported sociodemographic characteristics and reported membership of or sympathy with environmental groups are usually slight: what is reported is a tendency for some sociodemographic characteristics to be more frequently found in environmental campaigners, not a definitive relationship of cause and effect. For example, although having more education than the national average has been correlated with being more active in environmental campaigning, plenty of people who have more education than the national average are not more active environmentally and campaigners have *not* completed more education. So, sociodemographic measures offer only a partial explanation based on greater or lesser representation of certain groups in environmental publics.

Second, defining campaigning publics by how people self-report and self-identify in surveys does not reflect their practices. This means that different definitions of environmental campaigning can be covered by the same answer in a questionnaire, making interpretation problematic. Third, the pressure of social norms may also lead some people to report more environmental membership or activity than they are doing in practice, producing misleading data, whether intentionally or not. Any form of self-reported data can be problematic in this way, because it is usually unverifiable as well as influenced by hindsight and social expectations.

Some researchers have therefore attempted to define environmental activity more concretely by linking data on self-identification to data on actions. For example, in a 2000 survey, Dunlap and McCright (2008, p. 1056, Table 3) found that pro-environmental consuming practices, such as voluntarily recycling, using less water, reducing energy use and buying 'green' products, were reported by most people regardless of whether they self-identified with the environmental movement; on the other hand, more political and participatory behaviours, such as signing a petition or voting Green (see Chapter 7) were much more likely to be reported by those self-identifying with the environmental movement. In other words, people who think that they are part of a movement are more likely to say they practise environmental campaigning in the traditional sense of lobbying and working in groups.

Over time, the first set of consuming practices seems to have become far more mainstream than the second set of more traditionally political practices. Dalton (2015) said they had taken part in campaigning practices (such as belonging to an environmental group, signing an environmental petition or participating in an environmental demonstration) *less* frequently in 2010 than in 1993, but said they had taken part in domestic practices (such as recycling) and driving *more* frequently in 2010. Dalton interpreted this not as due to lifecycle or individual choice, but due to the institutionalisation of environmental action in legislation and regulation, especially recycling, that is, due not to what publics chose to do but to what they were asked to do by the state. He also found only weak correlations (around 0.18–0.25) between environmental campaigning and domestic bundles of other environmental practices.

So, we need to go beyond self-reporting to look in more detail at the practices performed by publics in environmental campaigning. To reiterate, we will focus not on the professional, institutionalised 'environmental movement' but on the practices that are taken up, adapted and shared by people who are campaigning on environmental issues more generally.

Reformist and radical, conventional and unconventional

One way in which campaigning practices have been analysed is using the labels 'reformist' and 'radical' to distinguish between campaigning practices or between groups that use them, as is commonly done in the literature (e.g. Doyle and McEachern 2007). Reformist groups are seen as less challenging to the status quo, working rather with existing elites and power structures than against them, and seeking change through small steps or partial improvements, that is, seeking to reform, rather than revolutionise, environmental policy and management. In contrast, radical groups are seen as explicitly challenging systems that they regard as unfair, unequal and oppressive, such as capitalism, patriarchy and authoritarian political regimes, seeking changes that are fundamental and widespread. For example, in the UK, the National Trust is seen as a reformist organisation, whereas Friends of the Earth was originally seen as radical in the 1970s but is increasingly seen as more reformist compared with the new radical groups that emerged in the 1990s such as Earth First!, Reclaim the Streets and Plane Stupid.

The kinds of practices that are usually labelled 'reformist' include signing petitions, organising meetings, attending public inquiries, seeking judicial review of planning decisions, fund-raising and recruiting new members. The kinds of practices that are usually labelled 'radical' include the bodily occupying of environmental sites threatened with destruction and damaging machinery that is poised to perform that destruction. Sometimes 'conventional' is used to refer to reformist campaigning using legal processes and 'unconventional' to refer to radical practices that involve illegal actions, criminal damage and/or civil disobedience as part of a politics of transgression. Groups using radical campaigning practices are usually seen as more anarchic, likely to be decentralised and more equitable in terms of decision-making, offering more personal autonomy to their participants in terms of choosing which topics they campaign on and how (Cotgrove 1982).

However, some groups mix and match practices. Greenpeace is seen by many people as radical in trying to block a large whaling ship with a tiny dinghy, but reformist in its hierarchical organisation that confines such actions to its core activists and asks 'ordinary' members to concentrate on fund-raising instead. Because of this, Jordan and Maloney (1997) argued that Greenpeace behaves more like a business than a campaigning group, despite blockading whaling ships not being standard business practice. Some radical groups also get more reformist with age, either because they change their practices and become more professionalised, or because new, more radical groups spring up a wave of grassroots mobilisation that make the older groups seem less radical by comparison.

The point I want to make for this chapter is that groups may practise both sorts of campaigning at different times and places, mixing and matching practices that could be labelled reformist or radical. In some cases, groups splinter precisely because activists disagree about what sort of practices will be more effective or appropriate or what division of labour a group will deploy across different sorts of practices. For example, Friends of the Earth was formed when David Brower left the (established and reformist) Sierra Club in the USA, in pursuit of more radical practices and greater orientation to grassroots members. As a result, local Friends of the Earth groups were able to set themselves up and choose their own targets for campaigning, while benefiting from the worldwide recognition of the Friends of the Earth name.

So the ephemeral and shifting memberships of groups and the practices and publics that enact them are continually being re-invented and re-valued, with the consequence that distinguishing between 'reformist' and 'radical' groups does not get us very far with understanding the campaigning practices open to and performed by environmental publics.

Practices in an environmental campaign

Instead, we need to think more about what kinds of campaigning practices make what kinds of environmental publics and vice versa. To do this, I will draw on an environmental campaign with which I was associated – the anti-incinerator campaign introduced in Chapter 3 and at the beginning of this chapter.

As well as the participatory practices already discussed in Chapter 3, the campaign involved organising and networking practices, such as setting up and running a 'grassroots' campaigning group, to represent people's concerns, to raise funds and to network with other campaigning groups and the public. Individual campaigners worked together to decide how to run their group, how closely it should work with other groups in the region, how its funds should be managed and what to call the group. Different people in the group sent emails and news-letters, gave speeches, debated the issues with representatives of the incinerator proposal in public meetings, set up websites, posted podcasts, produced maps, drawings and photomontages to counter those produced by the proposer, as well as raised funds through garden parties, charity quizzes and direct donations on websites and through personal contacts. The group also set up petitions on paper and online specifically to claim more-than-local opposition to the planning appli-cation, as well as to negate accusations of parochialism and NIMBYism (which we will come back to later).

These practices shaped environmental publics through building a group from individuals over time, drawing on practices from a recent (successful) campaign against an incinerator some miles away, as well as from the national umbrella organi-sation UKWIN. A core of a few individuals, mainly men, mainly white, some of whom had met before and some had not, officially became a group with a shared ethos and in-jokes, with other people being co-opted or temporary members in addition. The new group chose a name that was not geographically specific to the village or area, nor solely anti-incineration, thus deliberately shaping its own iden-tity and reflecting imaginaries of how campaigning groups are portrayed by the media in attempting to avoid being characterised as (merely) local objectors.

Their individual skills and capabilities were identified and exploited, including legal expertise from a solicitor, research and analytical skills from an academic, engi-neering and waste management expertise from those experienced in those indus-tries. Public speeches were rehearsed, with mutual feedback and support. Over six years of campaigning against the new waste facility, individuals invested not only their money and time but also their emotions, making and sometimes breaking friendships in the process. Meetings were held in village halls, in pubs and people's houses. Near the end of one campaign, the core campaigners met up for a final dinner, celebrating not a victory but the relationships that they had made through campaigning.

Some campaigning practices have adapted to online delivery (Van Laer and Van Aelst 2010). NGOs have for decades organised the mass-mailing of paper letters and pre-printed postcards to key targets such as government ministers, and such practices have translated readily to online versions, such as mass emailing and web-based peti-tions. Protest by internet or "cyberprotest" (Pickerill 2003) can be quicker and cheaper than traditional means, and mobilise resources for environmental campaigning by rais-ing awareness, recruiting new campaigners and organising events more quickly and adaptively, e.g. posting directions, times and places to meet online. Online campaigning can also make it easier for campaigning groups to network internationally and to mobilise global audiences to protest, e.g. at international summits of the G8 or about UN climate change policy (e.g. Van Laer and Van Aelst 2010).

But as well as translating existing practices into online spaces, some new practices have developed online, such as hacking corporate websites to access confidential information, to subvert their content or to damage their systems with computer viruses (Van Laer and Van Aelst 2010), although these have so far been little used in environmental cases. Van Laer and Van Aelst (2010) refer to "the virtual sit-in" as a form of "mass action hacktivism," where protesters coordinate to simultaneously send huge numbers of emails to a target, to try to overload and thus crash the target's internet server, disrupting the target's working practices as well as expressing the weight of feeling in terms of numbers of emails. Today, protesting often bundles together offline and online practices (e.g. Van Laer and Van Aelst 2010; Mercea 2012), e.g. using online social media to communicate information about and success of an embodied sit-in or rally.

As well as organising, campaigners develop diverse knowing and learning practices. In this local case, they did research and wrote and presented reports about highly complex specialised and technical aspects of proposals such as environmental pollution, waste technology and finance involved in building an incinerator. They also commissioned other people, such as environmental and engineering consultants, to do research and write reports to provide independent evidence for their campaign's arguments. Such practices aim to create more knowledge, to benefit from the authority that independent corroboration gives and to feed information to decision makers.

As time went on and the group learned more about the technical aspects of waste disposal, about the professional planners in the local authority who would influence the decision and about the elected councillors whose votes would make the final decision, the more the group felt that these officials lacked the specialised knowledge to understand the proposal and therefore to decide 'correctly'. Campaigners were often exasperated by what they saw as the lack of information displayed by officials and by political factors to do with the ruling party in the local authority outweighing objective and detailed information. They thus overturned the 'deficit model' of public understanding that we met in Chapter 2, in that these publics saw the powerful elites and 'experts' as misinformed, under-informed or even wilfully ignorant of the technical and economic details of the proposals. As self-made 'knowing publics', they sought to educate the non-expert councillors. Finding a university researcher specialising in waste management whose arguments about the problems of incineration resonated well with their own, they cited him extensively as an independent authority but also drew on their own considerable skills as solicitors, accountants, academics, engineers and company directors (also McClymont and O'Hare 2008; Bryant 1996) to analyse the data, choose legal strategies and manage their funds.

Through learning to be campaigners, environmental publics re-make themselves as 'knowing' publics, often casting those in power to be far less learned, as misinformed or under-informed and therefore as those who make 'bad' decisions, rather than as powerful experts. Emphasising practices is helpful here because it emphasises the regular re-enactment, development, sharing, collaboration and expansion that makes campaigns, rather than individual people or events. Again, it is worth emphasising that although I have separated types of practices for chapters of this book, in reality practices are far more mixed up and blurred.

As well as organising and learning, campaigning publics adopt, adapt and subvert legal practises to find loopholes for opposition and new ways to block or hamper environmental developments. In the incinerator case, campaigners spent many years working with(in) the legal system, by attending Planning Committees, requesting information from opponents using the UK's Freedom of Information Act and writing to the Secretary of State to demand a public inquiry on a local authority decision. Attending public inquiries or other meetings in person, demonstrating peacefully outside (e.g. Figure 6.1) or being inside in the public gallery all serve to show the strength of public objection through a mass of bodies. Campaigners may speak at inquiries or public meetings, to present evidence and persuade decision-makers to vote 'correctly' and to present alternative proposals, e.g. anaerobic digestion instead of incineration or a tunnel instead of a cutting for a new road in the case of the famous Twyford Down campaign in southern England (e.g. Bryant 1996). Performing such practices also enacts environmental publics within the legal system and through shared understandings of what constitutes 'good' conduct.

But sometimes publics deliberately seek to disrupt proceedings by misbehaving, deliberately going against the usual understandings of 'good' conduct. For example, a Twyford Down campaigner recalled with glee when talking to me in 1995 that their strategy to disrupt the 1976 public inquiry was to prevent it opening by making too much noise. Several hundred objectors attended in person and the Inspector was unable to make himself heard as they chanted 'No', sang hymns, rustled newspapers and hummed loudly, despite threats to remove them – "it was fun", he said.

Environmental publics are themselves also performed through these practices. Attending the 1985 public inquiry was a way for the campaigners against the Twyford Down cutting to express themselves within a legal setting, but in doing so they also met people with similar concerns and useful expertise whom they went on to work with (Bryant 1996, pp. 48–49). So the practices of attending, speaking and witnessing at the public inquiry also changed those publics as they were shaped, expanded and hardened by their experience. Indeed, Barbara Bryant (1996) talks of her life being taken over by the campaign to prevent the M3 cutting through Twyford Down, how her time, energies and money were all invested into it over years and years.

Illegal practices, direct action and civil disobedience

The diverse practices outlined so far depend upon writing, reading and discourse, and mainly worked within the law. However, some environmental campaigners move more deliberately to work outside the legal system, developing practices of putting campaigning bodies in the 'wrong' place to do the 'wrong' things, such as illegally occupying buildings or damaging machinery, not only using the body to physically represent the weight of public concern, but also making the body a key instrument of protest.

There is a long history of ordinary people using their own bodies in campaigns, e.g. seeking votes for women and civil rights in the 20th century. The mass trespass on Kinder Scout in the Peak District on 24 April 1932 used the physical presence of hikers to support the campaign for public access to private land, with six hikers being sent to jail. Bodily protest in the environmental cause became far more widespread later in the 20th century. In the 1970s, the embryonic Greenpeace group sailed tiny dinghies up to whaling ships, 'bearing witness' in the Quaker tradition by their bodily presence and making very good media visuals for news bulletins. In the 1980s and 1990s, environmental occupation and eco-sabotage were developed by Earth First! in various countries across North America and western Europe (e.g. Doherty et al. 2007; Seel 1997). In the UK, the Conservative government's massive road building programme prompted a range of anti-roads protests in the 1990s, which adopted, adapted and invented various forms of embodied campaigning practices.

Often referred to as 'non-violent direct action' (NVDA), these practices included occupying threatened sites with campaigners' bodies and causing damage to machinery such as road-diggers. In the 1990s, anti-road and anti-runway protesters in sites such as Twyford Down, London and Glasgow built precarious treehouses and tunnelled underground to make it more difficult for police and security personnel to remove them from proposed construction sites. Their enhanced vulnerability, swaying about at the top of a tree, made very good media coverage, which was one of its purposes, so that anti-roads protesters regularly made the front pages of national newspapers.

Illegal, disobedient and transgressive practices may end in arrest by the police, assault from security guards or both. Indeed, courting arrest has sometimes been a deliberate campaigning strategy, either as a badge of honour and proof of disrupting the (environmentally destructive) system or, more pragmatically, because arrests may attract more media attention than legal campaigning practices and also enable campaigners to articulate their protest when appearing in court some time later. For example, in 2008, Plane Stupid activists padlocked themselves to fences to shut down the main runway at Stansted airport for several hours, resulting in fifty-seven arrests but also national media coverage (e.g. *The Guardian* 8 December 2008).

And even when these practices have not actually been performed, the threat of them can still be conjured up in the imaginations of those involved in the same protest or of those involved in campaigns elsewhere. This means that the costs of policing and bad publicity can effectively travel ahead of such practices through storytelling, media coverage and networking:

> The threat of large-scale civil disobedience may therefore be much more effective in practice than the actual direct action campaigns themselves,

which invariably come too late in the day to make any difference to the completion of that particular scheme. Where they do make a difference is to the prospects for future schemes elsewhere in the country.[5]

Jonathon Porritt of Friends of the Earth, in Bryant 1996, p. 300

It is sometimes argued that NVDA and other illegal practices arise in response to the failure of more reformist approaches. McKay (1996, p. 128) quoted Emma Must, a Twyford Down campaigner in the UK, saying that "there have been 146 public inquiries into trunk roads and on only five occasions has the inspector found against the government. So it is no wonder that people end up in trees". In the USA, Earth First! was reportedly founded in 1980 "by a group of ex-reformist environmentalists who wanted to make the proposals of mainstream environmental organisations look reasonable in comparison" (Predelli 1995, p. 123). Allegedly inspired by Abbey's novel *The Monkey Wrench Gang*, the new group adopted practices rejected by established organisations such as driving metal spikes to deter loggers and other forms of eco-sabotage.

But in practice, environmental campaigns often involve both legal and illegal practices, with environmental publics adopting and adapting them differently across time and space. For example, when road construction began on Twyford Down in summer 1992, local campaigners, Winchester residents and those who had worked so hard to challenge the motorway through the legal system for years effectively accepted that they had 'lost' this particular battle. The local campaigning group, the Twyford Down Association, decided to stop fighting the battle through legal means, to continue giving advice and support to other campaigners but not to adopt illegal means or NVDA (Bryant 1996, p. 215). But others chose to do the opposite, setting up a protest camp on the threatened downland, and claiming a geographical identity by calling themselves the Donga tribe or Dongas, after the local name for the grassy ruts on Twyford Down made by driving livestock over the centuries.

Claiming and naming 'a tribe' was also a creative practice of making a public, an identity that diverse volunteers could share to bind them together and emphasise their common purpose, rather than their differences. Many of those in this Dongas tribe were new to environmental protest and thus "learned as they lived as they campaigned" (McKay 1996, p. 138): their living practices were also their campaigning practices. Their Dongas identity was expressed not only through a name, but also through styles of dress, music, food and slang. Such practices are sometimes interpreted as 'identity politics' or 'cultural politics' rather than traditional politics, because the personal experience of individual activists matters more than wider political ideologies or philosophies (e.g. Luke 1997, p. 29, referred to such protesters as both "post-Marxist" and "anti-Marxist"). Associated with 'New Age' travellers and "cultures of resistance" (McKay 1996), and tending to be younger, the Dongas seemed worlds away from the local, middle-class Conservative councillors, "the pearls-and-twinsets" (Bryant 1996, p. 192) who were also campaigning through the Twyford Down Association.

Such alliances of environmental publics can confound many assumptions about divides by age, class and political ideology, but also create problems. At Twyford Down, relationships between reformist and radical groups were uneasy and there were "accusations flying around of 'anarchists parachuting into the campaign' with no local knowledge and no local feeling" (Porritt, in Bryant 1996, p. 301). Barbara Bryant (1996) identified herself as a 'campaigner' rather than an 'activist', emphasising differences in self-identity as part of the 'tensions' between local residents and the New Age protesters.

Also, by their very nature, NVDA protest actions tend to be carried out by a small core of activists, even smaller than the core of established NGOs discussed earlier, maybe only ten to twenty activists in some cases (Doherty et al. 2007, p. 820). They often struggle to recruit new members outside the core clique: because the core is built through months or years of working together, shared experiences and friendship networks, when activists leave or 'burn out', new people are not easily recruited. Suspicion of newcomers can also prevent recruitment: many environmental campaigners[6] express fears of being infiltrated by establishment spies, fears that turned out to be well-founded when police undercover operations infiltrating environmental groups came to light much later, including some officers having sexual relationships with female protesters while using false identities (e.g. *The Guardian* 2014).

Perhaps because the campaigner's own body is at risk in campaigning, whether from injury or sexual exploitation, campaigning publics often develop their own particular styles of expression and tight bonds of friendship. But these may be short-lived and bundled together with other sorts of different practices, fading in and out as different tactics are tried, tested and dropped. And people move between groups, carrying practices with them or using new groups to generate new forms of campaigning. For example, one of the two people credited with founding Earth First! in the UK in 1991 later moved to work for Greenpeace UK (Bower and Torrance 2001): as well as groups, individual campaigners also shift and change.

Media, art and theatre in campaigning practices

A final set of practices has to do with generating the media coverage that is essential for environmental campaigns, raising awareness and recruiting more resources, bodies and knowledge to the cause. "Protest without media coverage is like a mime performance in the dark: possible but fairly pointless", suggests Jordan (1998, p. 327), so to achieve media coverage, protest groups and individuals send press releases to local radio stations, field spokespeople for interviews, write letters to local newspapers' letters pages and set up Twitter accounts, for example.

But there are many stories seeking coverage, so succeeding requires impactful visualisation, and an imaginative and theatrical politics of spectacle. This is not necessarily expensive: for example, before the 1985 public inquiry into the M3 cutting into Twyford Down opened, local campaigners used black polythene strips to mark out the sides of the proposed cutting across the hillslope, in a deliberately visual attempt to raise awareness and get media attention (Bryant 1996). Later, in 1992, protesters held a 'candlelit dinner' amidst the tents occupying the water meadows below the hill, with proper cutlery, chairs and a formal dress code, a peaceful protest that parodied English middle-class norms into a media-friendly photo opportunity. Shortly before the 1992 general election, the Twyford Down Association organised several hundred people to stand in the shape of a X on the white rock exposed by the initial stripping of soil from the top of the Down, producing a very media-friendly image (Bryant 1996, pp. 203–4) that symbolised not only voting on a ballot, but also the wrongness of the motorway cutting.

The politics of spectacle can thus be cheap, even for grassroots groups, and smartphones can upload videos of occupations immediately and circulate these widely using social media. Traditionally, campaigners have often self-filmed protests as a 'media subsidy', creating clips for journalists that they could import quickly and cheaply into their television reports, thereby giving campaigners some (but not total) control of how their campaign was portrayed, as well as increasing the chance of the campaign being covered by mass media. More recently, YouTube and similar internet services have made it easier to upload self-filmed and self-edited content directly for worldwide consumption, giving campaigners more control over how their campaigns are portrayed, what stories are told and which spokespeople are shown. As well as potentially reaching millions of people through the worldwide web, such coverage can be most influential when it is also picked up by mass media, e.g. television news.

This politics of spectacle can take different forms. Greenpeace have dressed up in tiger suits when protesting at Exxon Mobil shareholders' meetings, subverting the international advertising symbol of the company. The anti-motorway protests at Claremont Road in east London not only occupied threatened houses but also used artwork to transform a house due to be demolished into a gallery, with murals painted outside and objects displayed within, as part of a "festival of resistance" (Butler 1996, p. 341), including theatre, puppet shows and fancy dress parties that literally made a performance out of direct action (McKay 1996, p. 151). Creative practices reclaimed the space for countercultural protests, making the political also cultural, the destructive also creative and the confrontational also the carnivalesque, expressing environmental ideas through repertoires of humour, art, parody and subversion.

Even if campaigning spectacles fail to garner media coverage, protest as a performance can still have effects in terms of shaping environmental publics. Symbols and meanings are developed, subverted and parodied; spokespeople become more adept at packaging campaigning stories for media coverage; campaigners become film directors, advertisers and artists; people have fun and make friends.

Place

What does all of this diversity of campaigning practices suggest about the geographies of environmental publics? There are several points to make here, starting perhaps most obviously with the geographies of location, that is, where campaigns are based.

I have already detailed examples of campaigns to protect 'local' environments from the effects of building new incinerators, motorways and airport runways. Such campaigns often exploit and celebrate the particularities of place, both emphasising the value of the environments being defended and building identities for campaigning publics. The spatialisation of practice ranges from individual bodies walking across valued environments, such as Kinder Scout in 1932, to the construction of campaign landscapes through dwellings such as benders, treehouses, tipis and squats on threatened sites.

But protest also travels to be performed by campaigners in places far from the environment perceived to be threatened. For example, when Rootes (2000, p. 32) tried to map environmental protests based on reports in *The Guardian* national newspaper, he found that London was heavily over-represented, because parliament is based there and because the news media are more likely to report events on their doorstep than hundreds of miles away. This meant that national and international protests were travelling to London to make their protest more visible to a more powerful audience (government), as well as a large audience. For example, in 1990 Twyford Down campaigners were organising a demonstration in Winchester on the day that the minister for transport would announce the final decision about the Down, but their 'press mentor' asked them to move the demonstration to London, to enable them to be photographed on the doorstep of the Department for Transport and because the journalists would have to walk past them when leaving the minister's press conference, making it easy for them to be interviewed and thus included in television and radio coverage (Bryant 1996, p. 135).

And campaigning publics also travel. Local campaigners may initiate protests against new roads and airport runways, often beginning with conventional practices within the legal system, but non-local campaigners may join in later, bringing with them more confrontational practices and direct actions. When the second runway at Manchester airport was approved, the campaign moved from the local "middle class" residents and reformist practices of objecting to planning permissions and lobbying local authorities, to direct action practices of the bodily occupying of threatened environments by FoE, Earth First! and the Green Party (Griggs et al. 1998). When Friends of the Earth activists set up camp near Twyford Down in 1992 to occupy the site, local campaigners who lived nearby supplied hot food and opened their homes to campers who wanted to visit for a bath (Bryant 1996, p. 189).

Protest does not necessarily rely on bodily presence though. Activists have always shared practices through group networking, self-published magazines and training camps. The massive rise in internet use over recent years means that practices also travel virtually and very quickly, being shared through online and in-person relationships, through photocopied activist newsheets or photos on Instagram. The anti-road protests of the 1990s in the UK involved about 200 site-specific campaigns (Butler 1996) with their own practices and identities, but also shared practices and information through national networks such as Alarm UK. Diversity of practice thus adapted to (and was expressive of) diversity of place. As noted above, the Dongas tribe took their name from grassy ruts on Twyford Down outside Winchester, but that name rapidly become media shorthand for environmental occupations and direct actions at other locations too, where practices were similar.

Legal processes may also spatialize strategies of control. Some campaigners arrested for occupying threatened environments such as Twyford Down can be served with an injunction, a legal requirement not to return to the same site (e.g. Porritt, in Bryant 1996), a kind of 'spatial fix' through banishment. Although some campaigners may simply ignore the injunction and return to the protest site, others may not, meaning that more campaigners need to be recruited for campaigning to continue in the same site.

Other geographies of protest are not directly mappable in the cartographic sense, such as the spatial organisation of practices in public versus private spaces and the different restrictions and possibilities that these afford. Public spaces of environmental campaigning practice might be the public gallery of a town hall during a local authority planning decision (e.g. Figure 6.1) or a march along the streets of London ending in a rally on the green lawns outside parliament to protest against climate change. Private spaces might be the domestic space of the home where protest letters are written or the car from which people tweet campaign information, or the highly privatised spaces where shareholder meetings of big corporations take place and which activists seek to infiltrate and disrupt.

Scales of practice are also involved, as campaigning environmental publics consciously choose and enact their practices differently, depending on how they perceive scalar hierarchies of power and influence. This can be motivated by a group's perception that *other people* perceive the group as NIMBYs or perceive the environmental threat to be small in scale, lacking in interest or significance beyond local concerns. Most commonly, small and local groups seek to move their campaigns up a scale, because they perceive the national scale to be more powerful both in terms of national media coverage and in terms of national governments able to make decisions more quickly and even overturn local decisions. Practices include working with local MPs to persuade government ministers to 'call-in' a local planning decision to their own remit, seeking a Judicial Review of local decisions by appealing to the High Court, and appealing to supranational officials to challenge a decision made by national officials. Campaigning practices are therefore both embedded in the local but also often travel in order to network, to engage different audiences and to shift from tunnelling in the countryside to marching through the capital's streets, from legal challenges at the local town hall in northern England to legal challenges at the supranational level of the European Union in Brussels.

This favouring of higher scales of environmental issues, what we might call 'normative environmental scaling', often results in exhortations to environmental campaigns to 'upscale', that is, to move from the (merely) local to the national and international, in order to be more influential and effective by drawing on the greater discursive power associated with higher-order scales. Sometimes, this is referred to as 'jumping scale' or seeking a 'scalar fix' to problems, and usually implies changes in political and social structure of the organisations involved, not least that they must become more mobile or 'footloose' and less driven by grassroots concerns.

However, some groups also practise forms of 'downscaling' internally by cascading some power down through their own hierarchies, such as where local Friends of the Earth groups in the UK are able to choose what to campaign about as well as which other organisations to work with and what kind of information to draw on (also Rootes 2007), rather than have all campaign decisions taken by national headquarters. This can enable local publics to operate with some degree of autonomy in responding to their own concerns, while still benefiting from the support and resources of a much larger, nation-scale organisation. And even where local grassroots groups are not part of a larger organisation, they can still network in looser, more informal ways, with shifting alliances and often large turnover of participants from year to year who support a small core of committed, long-term campaigners. People thus move from 'attentive publics' to 'active publics' and back again, depending on the issues, their own lifecycle (how much time they perceived they have at their disposal), their perceptions of the likely success of campaigns and so on.

Similarly, campaigns for alternative food networks and against food miles and globalised commodity chains often reverse this normative scaling in order to value instead the small-scale, the local and the artisanal over the large-scale, the global and the mechanised production of food (Eden et al. 2008; Holloway and Kneafsey 2000). This reverses the hierarchical ordering of scales, but continues to deploy a scalar hierarchy as a shorthand that subsumes other social, political and economic attributes, prompting Winter (2003) to dub it "defensive localism".

So far, this chapter has drawn mainly on examples from the global North, especially the UK, but we can see international and global geographies of campaigning also. Earth First! originated in North America and spread mainly to western countries but less so to countries in the global South and East, perhaps because authoritarian governments there were more likely to quash the kinds of resistance and environmental protest that they practised. However, there are plenty of examples from the global South too. Probably the two best known are the Chipko movement in India and campaigning around the Amazon in Brazil.

The grassroots Chipko movement developed in the 1970s in Uttar Pradesh and, by the 1980s, it was renowned by academics and popular writers worldwide for mobilising environmental protest against the Indian government's policies for economic development. It was named for its initial bodily practice of 'tree hugging' to prevent logging by commercial contractors, a term which later became a pejorative shorthand for environmental idealists, similar to the way in which the term 'Dongas' was used to describe protests similar to those during the occupation of Twyford Down but in other locations.

However, the Chipko movement was not necessarily about tree preservation but about the rights of local communities (rather than commercial companies) to benefit from forest resources and to develop economically. Chipko gained global coverage "through simple, populist narratives that pitted peasants against the state and markets, but glossed over the heterogeneity of classes, interest, and constituencies"

(Rangan 1996, p. 216). The movement became a global inspiration for grassroots environmental campaigning because of the national forestry policies that it prompted, but when these further obstructed local communities' use of and economic benefit from forests in the region, resentment fermented so that, by the 1990s, Chipko had been overtaken by Uttaranchal, a movement for an independent state in the region (which was set up in 2000), showing that local development politics, rather than environmental politics, were the driving force here (Rangan 1996).

In Brazil, grassroots campaigning to protect rainforest in the Amazon basin from illegal and destructive logging has prompted worldwide coverage, which played on the Brazilian government's sensitivity to its negative image as the destroyer of rainforest, a sensitivity reinforced by media coverage of Rio de Janeiro's hosting of the UN environment summit in 1992 (Barbosa 2003). Partly in consequence, the number of environmental NGOs in Brazil grew from around forty to around 2,000 by the 1990s. Environmental campaigning was strongly linked to worker politics, making Chico Mendes globally famous as leader of the rubber-tappers in the 1980s. It was also linked to indigenous movements for civil rights using identity practices to claim 'Indianness' in the form of traditional clothing, body paints and traditional dances "as a political weapon" (ibid., p. 586) and claiming Indians as the guardians of the rainforest. This was not universal though: some Indian groups chose to support loggers and protested against the control of illegal logging (ibid., p. 587), emphasising that identity politics can be environmentally motivated or economically motivated.

Power

As well as being spatialized, environmental campaigning is riddled with assumptions about power. Asymmetries of power and worth are often applied to binaries like reformist/radical, local/national, grassroots/institutionalised, so let us take those in turn.

First, reformist groups are seen as failing to challenge the status quo, working instead with(in) existing structures of power for (merely) incremental improvements in environmental quality and protection through compromise and capitulation, whereas radical groups are valued as explicitly challenging power structures from the ground up, challenging oppression and inequality.

> We rightly admire the astonishing commitment of those who put their bodies where their beliefs are, in a succession of high-profile anti-road campaigns, but tend in the process to overlook the equally astonishing commitment of those who find it more fitting to do their bit in the backroom, in the Public Inquiry or even along the corridors of power.
>
> *Porritt, in Bryant 1996, p. 298*

In some ways, the definition of what is 'radical' changes over time, as the radicals become more accepted, more part of the establishment. It has also been argued that the origins of Earth First! and other direct action groups lie in reaction against the failures of established NGOs, but also have a strategic aim of making those established NGOs "look reasonable in comparison" (Predelli 1995, p. 123).

Second, the local/national binary is applied to power in very uneven ways. It implies a scalar hierarchy that often becomes shorthand for importance, with local concerns seen as less significant than national concerns. Sometimes local campaigns are seen as parochial but effective, raising the spectre of NIMBYism that we saw in Chapter 3 and prompting campaigning groups to try to present their interests as not site-specific but general (Gibson 2005, p. 393) in an attempt to upscale their campaigns into national and supranational chambers of power.

> The challenge for local campaigners is, therefore, to frame their grievances in such a way that their universality makes them into a political issue that transcends the normal and routine decision-making processes, that turns the issue from a routine planning dispute into a high profile political issue. This, of course, is a high risk strategy because to do so invites the big battalions of non-local political actors into the fray, making it likely that the contention will become one between non-local actors and reducing local campaigners to the status of onlookers.
>
> *Rootes 2007, p. 735*

Third, the grassroots/institutionalised binary is applied to power in order to hold up spontaneous campaigning by ordinary people as the democratic ideal of protest, emerging from the concerns of ordinary people, not from an elite removed from everyday life and/or pre-committed to a particular ideology, nor from the slick marketing and lobbying of established large pressure groups. Local campaigns are sometimes seen as more authentic, resonating with deeply felt local connections and spontaneous expressions of concern. Inglehart (1990, p. 63) argued that what makes 'new social movements' new is that they are not led by political elites, as political parties typically are, but by ordinary people mobilised by their own concerns, making them more worthy of support and legitimacy. Here it is not merely the geography of the local that is invoked, but also the closeness to ordinary people at the heart of democracy, prompted not by ideology or profit, but by strong connections to place and genuine feelings of concern for the future.

Jordan and Maloney (1997) took the opposite position by arguing that large environmental NGOs today behave like the old political parties: they are led by elites and recruit members not for political input but for income via subscriptions, legacies and sales of branded and environmentally friendly products, meaning that they do not mobilise environmental publics politically, but make them into merely consumers of environmental NGO ideologies and, by implication, the NGOs into brands. This argument pits consumption against citizenship in a moral ordering of campaigning practices, valuing publics who perform an actively political identity through campaigning above those who perform (merely) an economic one through consumption, even if that consumption is buying membership in a campaigning group. As we saw in Chapter 4, this is typical of the ways in which the figure of the consumer is brandished to denigrate consuming practices in comparison to others and to judge what is democratically worthy and what is not. Often this manoeuvre is motivated by the idea that democratically worthy practices are also more effective, but this is debatable, as we shall see when we look at voting practices in Chapter 7.

The expansion of social media in the 21st century has further complicated the geographies of the scalar hierarchy, as the geographies of internet protest can be argued to dis-embed practices from specific spaces and enable networking between NGOs across the globe. Maybe grassroots protests are becoming more "participatory" and potentially "subversive" (Pickerill 2003) by enabling anyone to send campaigning materials worldwide and to build networks that are no longer controlled by far-away or oppressive regimes. But while this may remove differences and obstacles based on geographical distance, it can introduce other differences and divides: as many people cannot participate in online social media if they do not have the income to buy the equipment, the skills and knowledge to use it effectively and the infrastructure to enable them to access the internet. This means that those who are poorer, older and living in remote areas poorly served by mobile reception and broadband are less likely to be enfranchised by the possibilities of cyberprotest.

Using social media is also often seen as a more democratic way to campaign over environmental issues, because it can get closer to the grassroots in that the content of social media is controlled by individuals who can directly influence how stories are told, which spokespeople get top billing and what images are included, unlike national television news coverage which is produced by professional staff in media corporations. Certainly social media has enabled millions of people to send their views around the world and to bypass national mass media, especially outlets controlled by authoritarian regimes. For example, thousands of people on the Chinese Weibo site (rather like Twitter) have microblogged about heavy air pollution in Chinese cities, especially Shanghai, contrasting the official but low readings released by the Chinese government with the higher readings released on Twitter by the US Embassy in Beijing since 2008, with the effect that "air pollution went from a taboo subject with data either not collected or shrouded in secrecy to a sensitive but widely discussed issue" (Kay et al. 2015, p. 354).[2]

But control is still exercised over social media, with some governments blocking, screening or deleting social media content, performing virtual censorship that may not be obvious to individual users. And as well as helping environmental campaigners network and communicate, social media can also help their opponents. For example, in 2013, the Chinese government monitored social media messages about the timing and nature of a planned protest at the site proposed for a new oil refinery, enabling it to send troops in to prevent the site being occupied by campaigners (Kay et al. 2015). Social media in this case enabled citizen activists to campaign, but also enabled the targets of their activism to listen in to their plans and more effectively counter or even prevent their direct action.

So, the geographies of environmental campaigning practices online may differ from those on the ground, but both involve gaps, obstacles and self-selection; rather than removing differences across time and space, online campaigning can generate them.

Summary

This chapter has deliberately not been about 'the environmental movement', but about how environmental publics perform various campaigning practices and, in the process, sometimes become part of (or even set up) environmental NGOs which, in turn, influence public values, knowledges and practices. It has also looked at how much 'the public' share or support the values and campaigning activities of established environmental groups, although this 'attentive public' proved difficult to measure precisely.

We saw that some environmental campaigning practices are developed by ordinary people in what are often referred to as grassroots actions, that is, activities outside the remit of established NGOs. However, these practices may still have been learnt from other NGOs – not necessarily environmental ones, as environmental activists have also learnt from the earlier peace movement, for instance – through networking in person or communicating online between groups. Environmental campaigning practices are very varied, from organising petitions, seeking judicial review of planning decisions, protesting at local authority offices, holding street parties and even occupying trees or tunnelling.

In contrast, some practices are developed by NGO elites and then directed towards or shared with a much wider public to try to mobilise mass support. A simple example is how environmental NGOs used to send out pre-printed postcards to their members, asking them to sign a postcard each and send them to a key official (e.g. a government minister) to express mass public support or opposition in the physical form of a bulging postbag. Social media has enabled this to be done even more quickly: mass emails can send hotlinks to online petitions to many people in an instant, although it has done little to speed up the legal processes involved in challenging decisions, such as judicial review. Sometimes the attentive environmental publics that are directed by NGO elites then circulate information further, such as when one person posts their intent to attend a demonstration to encourage all their Facebook 'friends' to attend as well, performing a further online mobilisation of others.

Other campaigning practices are explicitly learned through training or performed through sharing, such as when smaller, local protest groups consult larger, better-resourced national NGOs for advice, information and support. For example, Friends of the Earth supported local anti-road protesters at Twyford Down with advice and bodies to occupy the site, and in 1994 Greenpeace sent motorway protesters in Lancashire a road digger so that they could dig up the foundations being constructed for the road against which they were protesting (McKay 1996, p. 156).

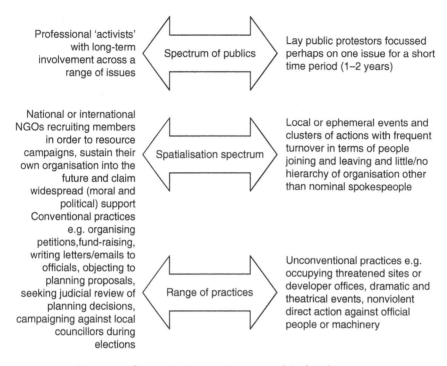

FIGURE 6.3 Spectrum of campaigning practices covered in this chapter.

To summarise these gradients in environmental campaigning publics, Figure 6.3 outlines the wide spectrum of possible practices and networking. A person might move from non-protesting to protesting outside a group, then to campaigning as a member of a group (or even set up a new group), maybe becoming a core activist or paid employee of a large NGO, or reversing these shifts to move back again to a non-activist role over time.

People make environmental campaigns through many different practices and those campaigns in turn make environmental publics of different sorts. Boundaries for campaigning are not clear-cut, however; rather they are hazy and shifting in time and space, especially as the weightings of scalar hierarchies of worth are both exploited by campaigning publics and used against them. Environmental campaigning is thus spatialized not only by location but also through repertoires of merit and influence, which both articulate and inform power relations more widely.

Notes

1   Instead of a pyramid, I could have drawn a set of concentric circles with the 'core' as the innermost one, which would have had the advantage of avoiding a pyramid's implication of hierarchy, putting the 'core' at the top. But it would have had the disadvantage of suggesting that publics revolve around the core, whereas for many people the core of environmental ideas is not at the centre of their worlds, but is of only occasional, perhaps fleeting, interest in their daily lives.

2   The International Social Survey Program 1993–2010, based on thirty countries in 2010 (including Eastern Europe, the former Soviet Union, Latin America and East Asia), twenty-four in 2000 and nineteen in 1993. In 1999, the highest reported was in the Netherlands at 45 per cent, nearly three times higher than the second highest (USA 16 per cent). Other countries reporting above 10 per cent were Denmark, Venezuela, Sweden, Greece, Belgium and Albania, with other countries tailing gradually down to nearly 0 per cent. These figures suggest large increases since 1981, when only the Netherlands reported over 10 per cent and the rest were all far lower (only eighteen countries were surveyed).

3   In 1986 the highest (50 per cent) reported was surprisingly from Greece and the lowest (8 per cent) from Belgium.

4   Dunlap (2010) noted that by 2010, although 19 per cent of those surveyed self-identified as an "active participant" in the environmental movement, only 42 per cent self-identified as "sympathetic" to it with 28 per cent "neutral" and 10 per cent "unsympathetic", suggesting a decline due to economic issues crowding out other concerns.

5   One reason is that those proposing schemes anticipated more protests in the same style and therefore more expensive security measures. For example, the Dongas' actions were estimated to add £3 million to the £26 million cost of the M3 cutting at Twyford (Bryant 1996, p. 308), making it 10 per cent more expensive.

6   Bryant (1996) was convinced her phone was tapped while she was campaigning against the M3 cutting at Twyford Down. Similarly, when I moved to Salisbury and tried to join the local Friends of the Earth group there, they were initially worried that I was a spy.

7   However, government and companies were the most frequent and influential bloggers, using the debate about air pollution outdoor to promote the purchase of air conditioners to combat air pollution indoors (Kay et al. 2015).

References

Barbosa, Luiz C. (2003). Save the rainforest! NGOs and grassroots organisations in the dialectics of Brazilian Amazonia. *International Social Science Journal* 55, 4, 583–591.

Bower, Jake and Jason Torrance (2 May 2001). Grey green. *The Guardian*, Society supplement, 8–9.

Bryant, Barbara (1996). *Twyford Down: roads, campaigning and environmental law*. E & FN Spon, London.

Butler, Beverley (1996). The tree, the tower and the shaman: the material culture of resistance of the M11 link roads protest of Wanstead and Leytonstone, London. *Journal of Material Culture* 1, 3, 337–363.

Cotgrove, Stephen (1982). *Catastrophe or Cornucopia: the environment, politics and the future.* John Wiley & Sons, New York.

Cracknell, Jon, Florence Miller, and Harriet Williams (2013). Passionate collaboration? Taking the pulse of the UK environmental sector. The Environmental Funders Network. www.greenfunders.org/wp-content/uploads/Passionate-Collaboration-Full-Report.pdf

Dalton, Russell J. (2005). The greening of the globe? Cross-national levels of environmental group membership. *Environmental Politics* 14, 4, 441–459.

Dalton, Russell J. (2015). Waxing or waning? The changing patterns of environmental activism, *Environmental Politics* 24, 4, 530–552.

Doherty, Brian, Alexandra Plows and Derek Wall (2007). Environmental direct action in Manchester, Oxford and North Wales: A protest event analysis. *Environmental Politics* 16, 5, 805–825.

Doyle, Timothy and Doug McEachern (2007). *Environment and Politics*, 3rd edition. Routledge, London.

Dunlap, Riley E. (2010). At 40, environmental movement endures, with less consensus. www.gallup.com/poll/127487/environmental-movement-endures-less-consensus.aspx

Dunlap, Riley E. and Aaron M. McCright (2008). Social movement identity: validating a measure of identification with the environmental movement. *Social Science Quarterly* 89, 5, 1045–1065.

Eden, Sally, Christopher Bear and Gordon Walker (2008a). Understanding and (dis)trusting food assurance schemes: consumer confidence and the "knowledge fix". *Journal of Rural Studies* 24, 1–14.

Gibson, Timothy A. (2005). NIMBY and the Civic Good. *City & Community* 4, 4, 381–401.

Greenpeace (2014). *Impact Report 2013.* www.greenpeace.org.uk/media/reports/impact-report-2013

Griggs, Steven, David Howarth and Brian Jacobs (1998). Second runway at Manchester. *Parliamentary Affairs* 51, 3, 358–369.

The Guardian (Thursday 21 August 2014). Undercover officers will not be charged over relationships with women. www.theguardian.com/uk-news/2014/aug/21/undercover-officers-relationships-women-not-charged

Holloway, Lewis and Moya Kneafsey (2000). Reading the space of the farmers' market: a preliminary investigation from the UK. *Sociologia Ruralis* 40, 3, 285–299.

Inglehart, Ron (1990). Values, ideology, and cognitive mobilization in new social movements. 44–66 in Russell J. Dalton and Manfred Kuechter (edited), *Challenging the Political Order: new social and political movements in Western democracies.* Polity, Cambridge.

Jancar-Webster, Barbara (1998). Environmental movement and social change in the transition countries. *Environmental Politics* 7, 1, 69–90.

Jordan, Grant (1998). Politics without parties: a growing trend? *Parliamentary Affairs* 51, 3, 314–328.

Jordan, Grant and William Maloney (1997). *The Protest Business: mobilizing campaign groups.* Manchester University Press, Manchester.

Kay, Samuel, Bo Zhao and Daniel Sui (2015). Can social media clear the air? A case study of the air pollution problem in Chinese cities. *The Professional Geographer* 67, 3, 351–363.

Lowe, Philip and Jane Goyder (1983). *Environmental Groups in Politics.* George Allen and Unwin, London.

Luke, Tim W (1997). *Ecocritiques: contesting the politics of nature, economy and culture.* University of Minnesota Press, Minneapolis.

McClymont, Katie and Paul O'Hare (2008). "We're not NIMBYs!" Contrasting local protest groups with idealised conceptions of sustainable communities. *Local Environment* 13, 4, 321–335.

McCormick, John (1992). *The Global Environmental Movement.* John Wiley & Sons Ltd, Chichester.

McKay, George (1996). *Senseless acts of beauty: cultures of resistance since the sixties.* Verso, London.

McNeish, Wallace (2000). The vitality of local protest: Alarm UK and the British anti-roads protest movement. 183–198 in Benjamin Seel, Matthew Paterson and Brian Doherty (edited), *Direct Action in British Environmentalism.* Routledge, London.

Mercea, Dan (2012). Digital prefigurative participation: the entwinement of online communication and offline participation in protest events. *New Media & Society* 14, 1, 153–169.

Nicholson, Max (1989). *The New Environmental Age.* Cambridge University Press, Cambridge.

Pickerill, Jenny (2003). *Cyberprotest: environmental activism online.* Manchester University Press, Manchester.

Predelli, Line Nyhagen (1995). Ideological conflict in the radical environmental group Earth First! *Environmental Politics* 4, 1, 123–129.

Rangan, Haripriya (1996). From Chipko to Uttaranchal: development, environment, and social protest in the Garhwal Himalayas, India. 205–226 in Richard Peet and Michael Watts (edited), *Liberation Ecologies: environment, development, social movements*. Routledge, London.

Rootes, Christopher (2000). Environmental protest in Britain 1988–1997. 25–61 in Benjamin Seel, Matthew Paterson and Brian Doherty (edited), *Direct Action in British Environmentalism*. Routledge, London.

Rootes, Christopher (2007). Acting locally: the character, contexts and significance of local environmental mobilisations. *Environmental Politics* 16, 5, 722–741.

Seel, Ben (1997). "If not you, then who?" Earth First! in the UK. *Environmental Politics* 6, 4, 172–179.

The Guardian (8 December, 2008). Climate activists held after Stanstead runway protest. www.theguardian.com/environment/2008/dec/08/stanstead-runway-protest.

Van Laer, Jeroen and Peter Van Aelst (2010). Internet and social movement action repertoires. *Information, Communication & Society* 13, 8, 1146–1171.

Winter, Michael (2003). Embeddedness, the new food economy and defensive localism. *Journal of Rural Studies* 19, 23–32.

7
VOTING PUBLICS

Introduction

The UK had one of the very earliest 'green' political parties (originally called People in 1973) and had been successful in local and even European elections for years, but it took nearly four decades for the 'green' vote to secure national political representation in the House of Commons. In the 2010 UK General Election, Caroline Lucas won the Brighton Pavilion seat for the Green Party with a majority of 1,252 votes. She was re-elected in the 2015 General Election, again as the only Green Party MP, with the Green Party gaining only 3.8 per cent of the vote across the UK.

Voting green is one very specific and reliably recorded environmental practice that is often used to characterise environmental publics. In this chapter, I want to think about those practices bundled up as 'green voting', not in order to analyse the environmental politics surrounding elections and Green Parties and their policies, but to consider what analysing green voting contributes to how we understand environmental publics. Not only do environmental voting practices vary across space and time, but those practices also make environmental publics in terms of our imaginaries of 'green voters', their relationships to power and how they use, reinvent and share tactics for making publics political.

This chapter is rather shorter than most of the rest in this book, mainly because the range of practices that count as 'voting' is far narrower than in other chapters, confined mainly to putting a cross on a piece of paper or pressing a button on a voting machine, although we will also consider belonging to political parties. At the same time, voting does not spontaneously appear and disappear in and of itself: it draws on and links to many other practices that we have already met, especially knowing, participating, campaigning and even consuming, so we shall also consider the links between practices and how far they are seen as more or less 'political'.

Practices of voting

Given those links, I could have chosen to bundle up voting practices into another chapter. I did not, mainly because voting is often seen as the quintessentially democratic practice that any public can perform and therefore seemed to merit special consideration. Why is that? First, it is valued because in many countries nearly every adult can do it, regardless of their income, disability, gender or race, at least where universal suffrage enfranchises everyone. By comparison, consuming, enjoying and other practices are certainly more constrained by an individual's income, disability or other personal characteristics. However, some groups are still excluded from voting, such as people in prison, immigrants without citizenship and children, depending on the electoral laws in different countries.

But more importantly, people who vote are often seen as more citizenly than those who do not, because voting is commonly understood to involve participating in a democratic process and acting properly on behalf of the civic good. Voting is therefore often held up as more political, more worthy for environmental publics to carry out, than other practices we have considered so far, especially more than consuming and enjoying, as we saw in Chapters 4 and 5.

But these ideals can collapse in practice, when we remember that many people vote based not on the civic good, but on self-interest, such as when issues like tax policies or anti-immigration controls are important to them. And many people do not vote at all, either because they do not legally register themselves to vote in the place where they live or because they are registered but do not exercise their right to vote by turning up to a polling station. In the 2015 UK General Election, for instance, only 66 per cent of registered voters actually voted and, in the 2014 European elections, the turnout across the twenty-eight member states varied from 13 per cent to 90 per cent, showing the importance of geographies of voting that we shall explore in more detail later.

So voting is a particularly circumscribed and controlled set of practices, designed by government expressly for publics to enact. Indeed, the expectation that citizens will fulfil their civil duty by voting is enforced by making voting compulsory in countries such as Austria, Argentina, Australia, Belgium, Bolivia, Brazil, Chile, Costa Rica, Cyprus, Dominican Republic, Ecuador, Egypt, Fiji, Gabon, Greece, Guatemala, Honduras, Italy, Liechtenstein, Luxembourg, Mexico, Nauru, Paraguay, Peru, Philippines, Singapore, Thailand, Turkey and Uruguay. Sanctions for not voting in different countries include not being able to claim state benefits or salaries, being fined, losing the right to vote in future, even imprisonment (Frankal 2005). In authoritarian and oppressive governments, however, the enactment of voting by publics may be more for show than for political influence, although in such regimes many other environmental practices to do with knowing, consuming and campaigning will also be heavily constrained or prohibited.

And the way in which people vote also varies. As we saw in Chapter 1, as well as practices involving meanings and knowledges, imaginaries and skills, practices use objects of all sorts. Traditionally, voting is done on paper with a cross made using a pencil, to include even those who cannot write anything else, but other practices are used across the world. The USA has used voting machines for some time to punch holes in a card to indicate voter preference, but problems with 'hanging chads' and other incompletely punched cards in the 2000 Presidential elections raised doubts about automatic counting when the final result proved very close. Many countries have also used indelible ink to mark voters' fingers after they have cast their votes, to prevent them (illegally) voting again in the same election. Voting practices are also highly controlled in terms of the timing of elections and the making of a mark by a citizen – ballot papers that are not filled in according to the electoral rules, whether accidentally or as a deliberate protest against the system, may be excluded by election officials as 'spoiled'. However, these processes and devices are less significant for environmentally influenced voting than the marked historical and geographical variations that we shall encounter later in this chapter.

Proportions

Let us now move on to consider what voting practices tell us about environmental publics. What proportion of publics vote because of environmental issues? What kind of publics are they? And how does this shape the public sphere more generally? The first question is fairly easy to answer because election results are often released in the form of detailed quantitative data, making it easy to compare geographical patterns, but the second is harder to answer and the third is extremely difficult. Let us take them in that order.

First, the proportion of publics who vote because of environmental issues can most easily be measured by looking at the number of votes gained by parties that self-identify as environmental. Usually lumped together as 'Greens', such parties use a variety of names depending on language and focus (Table 7.1). Green voting is a minority practice, where it is practised at all. For example, in the 2014 elections to the European Parliament, the green vote varied across twenty-eight countries from non-existent (in some countries, no parties contested the elections and self-identified as part of the Green/European Free Alliance grouping) to a maximum of about 15 per cent. In total, the 'green' political grouping won just over 7 per cent of votes cast internationally, sending forty-eight representatives to the European Parliament out of 751 seats contested, nearly double their result in the 1999 elections, when they won twenty-seven seats.

Another way to measure how many people practise environmental voting is to look at how many people join political parties that self-identify as environmental. Like the membership of environmental groups that we looked at in Chapter 6, such measures can be used as proxies to suggest the proportions supporting and sympathetic to green politics generally, although we need to remember that political parties of all colours tend to have much lower memberships than other sorts of voluntary, community or interest groups. It can also be difficult for researchers to obtain figures for party membership, as not all parties declare them, but Table 7.2 uses figures available or estimated for UK political parties to show that the memberships of Green parties

TABLE 7.1 Green/European Free Alliance voting in the 2014 European elections

	% votes cast	Green/EFA MEPs elected	Party names
Austria	14.5	3	The Green Alternative
Belgium	6.7	1	Groen!
	0.1	1	Ecolo
Bulgaria	–	–	
Croatia	9.4	1	Croatian Sustainable Development (ORaH)
Cyprus	–		
Czech Republic	–		
Denmark	11	1	Socialist People's Party (SF)
Estonia	13.2	1	Independent Canididate (Indrek Tarand)
Finland	9.3	1	Green League (Vihr)
France	8.9	6	Europe Ecology – The Greens (EELV)
Germany	10.7	11	Alliance '90/The Greens (Grüne)Pirate Party
	1.4	1	
Greece	–		
Hungary	7.25	1	Together 2014
		1	Politics Can Be Different (LMP)
Ireland	–		
Italy	–		
Latvia	6.4	1	For Human Rights in United Latvia (PCTVL)
Lithuania	–		
Luxembourg	15	1	The Greens
Malta	–		
Netherlands	6.9	2	GreenLeft (GL)
Poland	–		
Portugal	–		
Romania	–		
Slovakia	–		
Slovenia	–		
Spain	10	6	United Left (IU)
	4	2	Left for the Right to Decide (EPDD) – Coalition of Republican Left of Catalonia (ERC) + New Left of
	2.1	1	Catalonia (NECat)
	1.9	1	The Peoples Decide (LPD)
			European Spring (PE)
Sweden	15.4	4	The Green Party Green Party
United Kingdom	7.6	3	The Greens
	2.4	2	SNP (Scottish National Party)
	0.7	1	Plaid Cymru (Welsh Nationals)

Note: Country results are reported in House of Commons Research Paper 14/32 11 June 2014: European Parliament Elections 2014. Only parties that are within the Green/European Free Alliance grouping are included here.

TABLE 7.2 Membership, votes and Members of Parliament of UK political parties gaining representation in Parliament in the 2015 General Election, ranked by number of MPs

	Party members*	Votes gained	Ratio of members to votes	% of vote 2015	MPs elected
Conservative Party	149,800	11,334,576	1:76	36.9	331
Labour Party	190,000	9,347,304	1:49	30.4	232
Scottish National Party	93,000	1,454,436	1:16	4.7	56
Liberal Democrat Party	44,000	2,415,862	1:55	7.9	8
Democratic Unionist Party		184,260		0.6	8
Sinn Féin		176,232		0.6	4
Plaid Cymru	Estimated at nearly 8,000	181,704	1:23	0.6	3
Social Democratic & Labour Party		99,809		0.3	3
Ulster Unionist Party		114,935		0.4	2
Green Party (England and Wales) and Scottish Green Party Green Party	44,000**	1,157,613	1:26	3.8	1
UKIP (UK Independence Party)	42,000	3,881,099	1:92	12.6	1

Notes:
* as of January 2015.
** Membership figure based on the UK Green Parties coalition, within which the Green Party (England and Wales had around 35,500 members and the Scottish Green Party around 8,000). Sources: Keen (2015, for House of Commons) and BBC (2015). Turnout was 66.1 per cent overall, and the total number of people able to vote was over 46 million.

are typically low as a percentage of the electorate, but so are those for other political parties.[1] Also, the number of members in a political party does not easily translate into the number of votes that party gains in an election, with the Green Party in the UK gaining fewer votes in proportion to their number of members than other parties, especially those parties towards the right of the political spectrum.

It is not only by voting green that publics can practise environmental voting: other parties may have environmental policies that influence people's voting practices, especially where they feel strong allegiance to a mainstream party but are also concerned about environmental damage. But this influence also tends to be low or concentrated in a minority of the electorate. For example, in the 2001 British General Election, the mainstream parties had little in the way of environmental policies in their manifestos: Carter et al. (2001) considered Labour's environmental stance as "low-key" and "weak" and the Conservatives as "distinctly anti-environmental". The Labour Party won the election, suggesting that environmental issues played little part in voting patterns.

My second question was, what kind of publics vote green, that is, do green voters share common characteristics or show particular attributes? Political scientists in particular have attempted to answer this question for decades, using the large datasets of voting results as well as pre- and post-election surveys of voters. This can be problematic when surveys rely on self-reported attitudes and voting patterns, rather than actual voting (which usually remains secret), and thus may be inaccurate or misleading; studies have nevertheless attempted to use these data to profile 'green voters' by various sociodemographic attributes.

Age is one such characteristics and young people (and students) have been found to be more likely to vote green, e.g. in the UK in the 1989 European elections (Rüdig and Franklin 1992; Franklin and Rüdig 1995) and more recently in Australia (Tranter 2012). In Germany, green voting by younger people first put the Greens into the parliament in 1983, when more than two-thirds of green voters were reportedly under 35 years old and one-third were under 25 years old. But as Green supporters aged, so green voting moved into older age groups and by 2009, green voters in Germany presented a mixed picture of young and first-time voters as well as older voters. Rüdig (2012, p. 117) suggested this was due to the ageing of the cohort that first protested, perhaps as students, during the 1968 counter-culture movements, including the environmental movement. Chapter 6 looked more explicitly at such protest practices, but it is worth emphasising here that voting practices are distinct but not separate from other sorts of environmental practices: environmental publics may practise voting, protesting, consuming, participating and knowing in the same times and spaces, maybe on the very same day, so that practices are shared, mutually influential, but also sometimes seemingly contradictory.

As well as age, other sociodemographic characteristics reportedly correlated with (self-reported) green voting in various studies are: being female (Kellner 2009; Rüdig 2012), being in higher social grades or middle-class groupings (Kellner 2009; Tranter 2012) and having completed more formal education (Kellner 2009; Tranter 2012). Those reporting voting green also tend to report more liberal and leftist political views and affiliations (e.g. Dalton 2015; Dunlap and McCright 2008; Kellner 2009).

However, survey results are often mixed, rather than clear-cut. Younger people may be *more likely* to vote green than older people, but by no means do all younger people do so, and many older people also vote green. Green voters therefore do not present as a clearly defined public and the only common attribute they seem to share is, perhaps obviously, a sense of environmental concern. Dalton (2005) also found that those who reported voting green in national elections in 2000 were also more likely to report that they were members of environmental groups, suggesting that green voting is linked to green campaigning, as discussed in Chapter 6. The size of the green vote also corresponds reasonably well with the size of environmental group membership, at around 5–10 per cent of the population in many western countries. There is potential for it to grow by reaching more of what I referred to in Chapter 6 as the 'attentive' public, which covers the majority of people.

Now to my third question: how does Green voting (or at least the possibility of green voting offered by the presence of self-identified Green parties or candidates) shape the public sphere more generally? One argument is that when votes are cast for Green parties, even if their candidates do not win, this still sends a message to the other (non-Green) parties that environmental policies are potential vote-winners and may influence them to develop the environmental aspects of their own manifestos more. I already noted that this has not happened much in the UK and, with the global economic recession since 2008, there seems little change in this so far.

In some cases, Green candidates may gain enough support to win seats and thus enter the official political decision-making elite, taking into government the environmental promises that they made to their publics and thus enabling green voting to directly influence the public sphere more generally. However, this is rare in many countries at the national scale; Germany is the main exception where the Greens are far more prominent in government, having first entered parliament in 1983 (see Rüdig 2006).

Place

This leads us onto the question of the geographies of environmental voting, which are both important and well documented using electoral data. To start with, it is perhaps obvious to say that voting is a very explicitly spatialized practice compared to others we have looked at. For example, I can sit in my house in northern England and buy a product from across the world on eBay or travel across the world with a plane ticket, but when voting, I am restricted to my own country in general elections or my own council (or parish) area in local elections, even if I move to live somewhere else on election day. I can only vote in one place in person, embedding me in a particular electoral infrastructure, and I cannot transfer to another without considerable effort, although I can vote remotely through the post from wherever I have chosen to travel.

And voting is spatialized not only by place (my residential address, for example) but also by scale. I can vote in local elections, national elections and even supranational elections, such as those to elect representatives to the European Parliament, and the voting practices in each case often differ greatly, as I shall show below. Many countries (and individual US states) also run referenda on controversial questions of policy, governance and their constitutions, in which voters do not elect candidates to powerful roles, but do effectively make political decisions that governments are required to enact.

So, in terms of environmental publics, the scalar organisation of voting and the political campaigns that lead up to it are particularly pronounced. Let us take some examples.

Voting locally

We can start with voting practices scaled as 'local'. Of course, in one sense all voting is local because the ability to vote usually depends on one's registered address and legal status as a citizen of a nation-state, that is, the geography of residence. But voting practices are also scaled, that is, they are differently organised and enacted according to whether they are set up and perceived to be for local government, national government or even supranational government in the case of the European Union. The scaling of voting is thus part of how voting practices are understood and organised, with voting practices also influenced by how those publics perceive these scales and in particular the relative worth or power attributed to different electoral processes.

Let us look at 'local' elections in which publics vote for candidates aspiring to roles in local and regional government, whether for local authority councillors as in the UK or for government officials such as sheriffs and attorneys, as in the USA. In the UK, green councillors were elected to local government roles decades before the first Green Member of Parliament (MP) was elected to the national government in 2010. For example, in the 1999 local elections held at the same time as the European elections, the Greens won 11 per cent of the votes for the Greater London Assembly, resulting in three representatives; they also gained one Member of the Scottish Parliament and forty-four local councillors. But they won only 6.3 per cent of the UK vote for the European parliament in 1999, resulting in two green MEPs going to Brussels, and two years later in the 2001 General Election to the national parliament in London. Carter et al. (2001, p. 103) called the Green Party "a mere sideshow" for gaining only 0.7 per cent of the vote (166,487 votes). By January 2015, the Green Party had more than 160 principal authority councillors in England and Wales (The Green Party 2015), but still only one MP.

In the USA, the Green Party fields candidates in national elections, most famously Ralph Nader running for President in 1996 and 2000 (when he was backed by the Green Party, see Bomberg 2001), as well as many candidates for the House of Representatives and the Senate. So far, they have had little success, because the two-party system is firmly entrenched and huge budgets are necessary for national advertising and campaigning. In the 2012 election for President, for example, the Green candidate was Jill Stein and she came fourth nationally with only 0.36 per cent of the vote (469,627 votes, Federal Commission 2013), although this varied from nearly 0 to over 2 per cent in different states. Candidates for senate seats similarly gained from nearly 0 to nearly 9 per cent of the vote in different places (the highest being in New York state). As in the UK, Greens have been more successful in local elections in the USA, not only for local government councillors (as is common in other nation-states), but also for mayors, judges, sheriffs, attorneys, auditors and even coroners in different cities, counties and states across the country (Green Party of the United States 2015).

So when green voting is locally scaled, it often shows more potential for those individuals choosing to stand for public office with a green affiliation and those members of the public choosing to vote because of green affiliations, particularly in comparison to nationally scaled elections, which we turn to next.

Voting nationally

Geographical patterns of voting reflect how the voting public views the scale and importance of different elections. As noted in the introduction to this chapter, the UK has so far only elected one green representative to the national parliament – Caroline Lucas, first elected as an MP in 2010 and re-elected in 2015. Voting in the UK is especially spatialized because of the 'first past the post' electoral system, in which each constituency is geographically bounded as a single unit that is won by the candidate with the most votes. In the UK, the Greens do better in local and European elections than in national elections. In the 2009 European elections, the Greens in the UK did well to gain 9 per cent of the vote, albeit with a low overall turnout of one-third of the electorate, but in the following year's national election, although the first Green MP was elected, environmental issues seemed to have little influence on other voting patterns.

We can compare this to Germany, where the Greens gained national representation much earlier and have secured far more representatives in positions of power nationally than in the UK. Despite this, the collapse of the green vote in Germany has been predicted frequently, especially whenever economic conditions worsen. In the 2000s, the German Greens were expected to suffer a loss of support because of the global recession, but also because of negative media coverage of Red/Green ideological divisions within the party since 2005 and because of losing Joschka Fischer, who was foreign minister for 1998–2005 and reportedly the "most popular politician" not only in the Green Party Green Party, but also in Germany[2] (Rüdig 2012, p. 113). Yet the Greens continued to do well after 2005, gaining 12 per cent in European Parliament elections in 2009, over 10 per cent in national (Bundestag) elections, 12 per cent in 2010 regional elections in North Rhine-Westphalia, 24 per cent in 2011 regional elections in Baden-Württemberg and their first regional prime minister (at the head of a Green-Social Democratic Party coalition) with over 17 per cent in the Berlin mayoral election (Rüdig 2012).

In Greece, the Greens emerged much later, only in the 2000s. They surprised analysts by gaining 3.5 per cent in the 2009 European Parliament election and their first MEP, then 2.5 per cent in the national election by attracting "disenchanted and cynical voters who opposed all parts of the three-pole system" of Left/Centre/Right that was well-established in Greece (Vasilopoulos and Demertzis 2013, p. 731). Green votes were initially perceived as 'protest' votes against mainstream parties, as well as prompted by concern over environmental problems such as an outbreak of wildfires in 2007, shortly before an election. But as opposition amongst the Greek electorate to austerity measures imposed by the EU and fears that Greece would leave the EU and the euro both grew, by 2012 Green voting in Greece had faded away, with no Green Party participating in the 2014 European elections in Greece (see Table 7.1).

So the geographies of voting practices matter here, shaping the impact of Green parties and policies but also shaping how voting publics think of themselves. Often, voting Green is seen as a 'protest vote', as in Greece, especially in elections that publics see as less important (such as local or European ones), and also when fewer people turn out. For example, in the 2009 elections to the European Parliament, the Greens in the UK did well to gain 9 per cent of the vote, but other smaller parties without sitting MPs also did very well, with the UK Independence Party gaining 17 per cent of the vote and the British National Party gaining 6 per cent. In total, those parties *without* sitting MPs gained 40 per cent of all UK votes, suggesting that 'protest' voting was important here, as well as the low overall turnout of only one-third of the electorate.

Voting internationally: the case of Europe

As we have already seen, European elections are a special case of supranational voting across member states in a world region where Green parties first developed and where they have been most successful in terms of elections, mainly at local levels. But even within this scale, there are important geographies of green voting, in terms of differences between the member states of the European Union (EU) and in terms of differences between elections to the European Parliament and elections to national and local parliaments. Let us take those two in turn.

First, the Green vote varies greatly between the member states of the EU, as does the history and strength of Green parties. As seen in Table 7.1, Greens did well in the 2014 elections to the European Parliament, achieving over 10 per cent of national votes in Austria, Denmark, Estonia, Germany, Luxembourg, Spain and Sweden. However, the Greens' percentage of the vote did not translate into Green MEPs in a straightforward way, e.g. compare Austria where the Greens won over 14 per cent of the vote and sent three MEPs to Brussels with Croatia where they won over 9 per cent of the vote but only sent one.

Overall in 2014, the 'Green' political grouping won over 7 per cent of votes cast, gaining 48 Members of the European Parliament (MEPs) out of 751 seats contested, nearly double their result in the 1999 elections, when they gained 27 MEPs. In the UK, the Green Party of England and Wales came fourth nationally in 2014 with 1,255,573 votes (nearly 8 per cent) and three MEPs elected out of the seventy-three MEPs representing the UK. Drilling down into the regional pattern in more detail shows that the Greens gained one new MEP in the southwest of England, where the Green vote peaked at just over 11 per cent, and retained their two MEPs in London and the southeast of England. However, outside the southwest, the green vote in other regions was below 10 per cent and lowest in Wales at 4.5 per cent (see BBC 2014).

The largest scale of elections, the supranational, is also the most complex and uncommon, but shows that environmental effects are influenced both by place – the geographical location of voting – but also by scale – the perception that some elections are worth more than others. Voting Green as a protest against mainstream parties is thus more frequent in elections that are seen as less important, despite their large scale.

Power

Finally, what do these patterns of environmental voting suggest for the power of voting publics? First, we need to remember that theories of citizenship as power often assume that voting practices are more powerful than buying or other domestic practices of using, recycling, reusing, because voting can influence the state, that is, the most powerful individual agent of environmental action that can be imagined outside of natural forces. This echoes the binary we met in Chapters 4 and 6 between campaigning and consuming publics, a binary that is used specially to denigrate consuming practices as less morally worthy, less effective and more delusional than other participatory practices such as voting or campaigning. This argument revolves not only around what people do, but also around how far what they do counts as political or not. It is easier to regard voting as a traditionally political act by ordinary people than it is to regard choosing what kind of oranges to buy as political, because the choice to buy is more commonly seen as monetarised and dependent on personal preference rather than the civic good.

Pitting 'consumers' against 'citizens' in a judgement of moral worth reflects hundreds of years of dualistic understandings in which consumers are negatively seen as acquisitive, materialistic, superficial, individualised, wasteful, inauthentic and unpolitical as they live their selfish lives in the private sphere, whereas citizens are positively seen as embracing their democratic role and supporting the public good by participating in the public sphere of civic action through active campaigning, canvassing and voting. Yet what is rarely pointed out is how much the latter conceptualisation of 'the citizen' is itself an idealisation, with many citizens also committing crimes, refusing to pay taxes and living selfish lives, not least by voting in self-interest and failing to engage with others in community projects.

This binary of citizen/consumer therefore works to diminish the worth of some practices over others in a way that Soper and Trentman (2008, p. 4) argued is restrictive rather than useful. The problem is that such arguments are about *identities*, not *practices*. A moment's reflection will suggest that the same people who practise consumption also practise citizenship, because the same person who buys paper in a supermarket may vote in an election, maybe on the same day. But voting is still often seen as the quintessentially citizenly practice – a democratic right hard-fought-for in many countries and by many disadvantaged and oppressed groups, such as women and people of colour. Voting is therefore often held up as more worthy, more unquestionably political, more altruistic than other practices of consuming or enjoying.

This attitude persists despite low numbers often turning out to vote in many elections, showing clearly that many people do not exercise their democratic right by voting, while many more may exercise their monetary power by buying things.

For example, from 1945 to 2015 there were nineteen UK general elections in which turnout varied from a high of 83.9 per cent in 1950 to a low of 59.4 per cent in 2001, averaging 73.5 per cent over the period (UK Political Info 2015). In Europe, turnout varies hugely across member states, with turnout in the 2014 European Parliament election ranging from a high of 90 per cent in Belgium to a low of 13 per cent in Slovakia, with an average of under half of the electorate (see Table 7.3).

In the USA (see Table 7.4), the voter turnout for presidential elections every four years from 1976 to 2012 has been only just over half of all eligible registered voters nationally, with a high of 57 per cent in 2008, when Barack Obama won, and a low of 49 per cent in 1996, when Bill Clinton won. So nearly half of all eligible voters in the USA regularly do not vote. In the mid-term Senate elections in 2014, when there were no simultaneous Presidential elections at the national scale, nearly two-thirds of registered US voters did not vote, suggesting that voting publics in the USA also see the national scale as more important than the regional or state scale.

So maybe the citizen/consumer argument should be turned around to politicise ordinary consumption more as "a vital source for political action" (Soper and Trentman 2008, p. 11), because, in many countries, more people practise consumption than practise voting. Indeed, nearly everyone in developed and urbanised countries buys things but in those same countries many people – in some elections, *most* people – do not vote. If more people buy things than vote, if more people belong to the National Trust to consume the lovely views of/from the land that it owns than belong to political parties, then maybe we ought to analyse why consuming is more important than voting in terms of political expression by environmental publics.

Having said that, it remains difficult to separate out strictly environmental reasons for voting. In the same way that some people may buy (and use) products because of environmental factors, but others may buy and use the same things because of non-environmental factors (especially cost), some people may vote Green because of environmental factors but others may do so due to non-environmental factors.

As noted, a vote for a Green Party (or other small party) is often seen as a 'protest' vote against the mainstream parties. In the 1989 European elections, the UK green vote of 14.9 per cent was unprecedented, having reached only 0.5 per cent in the previous European elections in 1984, and greater than in West Germany (8.4 per cent) or France (10.6 per cent), where green voting was more established (Rüdig and Franklin 1992). As a result, the UK green vote was widely interpreted as a "protest" vote (Franklin and Rüdig 1995) or "experimental" vote (Rüdig et al. 1996) against the Conservative government then in power; it was argued that the UK voting public felt European elections were unimportant compared to national ones but did offer the voting public an opportunity to express dissatisfaction with the 'traditional' parties without significant consequences; voters felt enabled to send a political message without cost. In subsequent elections to the national parliament, the green vote in the UK declined, suggesting that the 1989 green vote came mainly from 'floating' voters, that is, they voted green not in a repeated and committed way, but as an irregular and sporadic practice.

TABLE 7.3 Turnout (per cent of voters who voted) in the 2014 European Parliament elections across EU member states, ranked

Belgium	90%	Netherlands	37%
Luxembourg	86%	Estonia	37%
Malta	75%	Bulgaria	36%
Greece	60%	United Kingdom	36%
Italy	57%	Portugal	33%
Denmark	56%	Romania	32%
Ireland	52%	Latvia	30%
Sweden	51%	Hungary	29%
Germany	48%	Croatia	25%
Lithuania	47%	Poland	24%
Austria	45%	Slovenia	24%
Cyprus	44%	Czech Republic	18%
Spain	43%	Slovakia	13%
France	42%		
Finland	39%	*Average for EU*	*43%*

Notes: 2014 figures taken from UK Political Info (2015) and rounded up to nearest per cent.

TABLE 7.4 Turnout (per cent of eligible voters who voted) in US elections for President and House of Representatives since 1978

	Turnout in Presidential elections as % voting-age population	Turnout in House of Representatives elections as % voting-age population
1976	53.6	48.8
1978		34.5
1980	52.8	47.5
1982		37.7
1984	53.3	47.4
1986		33.6
1988	50.3	44.9
1990		33.6
1992	55.2	51.3
1994		36.5
1996	49.0	45.9
1998		33.1
2000	50.3	47.1
2002		34.8
2004	55.7	51.6
2006		36.1
2008	57.1	53.3
2010		37.0
2012	54.9	56.5
2014		38.5

Source: Data drawn from U.S. Census Bureau (2012, 2014) and US Federal Commission (2013).
Note: Presidential elections are held every four years across all states, but House of Representatives elections are held every two years.

This was tested in a post-election survey of voters (Rüdig and Franklin 1992, p. 42), which found that only 42 per cent of those who reported voting Green in the UK in 1989 said that they would vote Green in a national election. In comparison, 95 per cent of those who reported voting Conservative (the party in government at the time) said that they would vote Conservative in a national election and 96 per cent of those who reported voting Labour (the mainstream opposition party at the time) said that they would vote Labour in a national election. This suggested that green voting did not present much of a long-term threat or green pressure on the mainstream parties.

In Greece, the Green vote in the 2000s was considered to be a protest vote against *all* the mainstream parties, due to widespread scepticism towards the party system as a whole. This resulted in what Vasilopoulos and Demertzis (2013, p. 735) called a "negative vote for the Greens" that exceeded that of any of the other major parties. Yet by 2014, this Green vote had faded away and the Green Party did not even contest the Greek elections to the European parliament in that year.

Such contradictions and shifts again emphasise the need to think in terms of practices, not identity. People may join Green political parties because they think of themselves as 'green' voters, but many voters choose to vote Green in one election but not in another, even when the elections are happening at the same time. And in some cases, voters may support candidates or parties that are not explicitly 'Green' but still have environmental policies, e.g. the Liberal Democrats in the UK used to be seen as the greenest of the mainstream parties, as well as the most Europe-facing.

Tactical green voting is another possible practice. Campaigning groups have recommended that people use 'tactical voting' in local and general elections to try to get candidates elected who have promised support for particular environmental campaigns, even if those candidates are not standing for an explicitly Green Party. For example, in the Twyford Down protests that we met in Chapter 6, local campaigners chose to support non-Conservative Party candidates standing in the 1992 General Election in constituencies covering and surrounding Twyford Down, a region traditionally dominated by Conservative support. They leafletted local publics to urge them to vote for the chosen candidates rather than for the Conservative Party candidates, because the latter were (implicitly) supporting environmental damage through supporting the national Conservative government's road-building programme.

Finally, when considering how 'green' voting practices are linked to the exercise of citizen power and the making of environmental publics, there are three questions to address. First, does voting 'green' have influence in the sense of electing green(er) governments, making environmental publics performative not only of their own lives but of the ongoing development of the state? So far, there is little evidence for this except in Germany and a few other western European nations where Green Party candidates have been elected to positions of parliamentary power.

Second, does voting 'green' have influence in the sense of persuading mainstream parties to adopt or support environmental policies? As I have suggested, Green voting practices are not restricted to voting for Green parties, but may include voting for non-Green parties for environmental reasons, such as concerns about environmental damage, renewable energy or air pollution. To put this another way, green voting is not a single practice but a bundle of related practices that include voting for parties but also voting against them. But again there is little conclusive evidence of influence through this route.

Third, does voting 'green' reinforce other sorts of practices such as consumption, that is, how does voting link with or arise from other relationships and practices through which people perform their daily lives? In the U.S. Gallup poll in 2000 analysed by Dunlap and McCright (2008, discussed in Chapter 6):

- 27.5% of those surveyed said that environmental matters influenced the way they had voted in an election (of any type)
- 42.3% of those self-identifying as 'active' in the environmental movement reported voting environmentally
- 29.1% of those self-identifying as 'sympathetic' to the environmental movement reported voting environmentally,
- 17.8% of those self-identifying as 'neutral' to environmental movement reported voting environmentally and
- 8.3% of those self-identifying as 'unsympathetic' to environmental movement reported voting environmentally (Dunlap and McCright 2008).

Far more people reported consuming more environmentally by recycling (90.3 per cent total), conserving water (81.95 per cent total) and conserving energy (83.3 per cent total), than reported voting environmentally, suggesting that environmental voting practices link only loosely with other sorts of environmental practices, but also that they are far less prevalent amongst most publics than those other practices.

Summing up

Voting publics offer some interesting contradictions. Voting is often seen as the quintessentially democratic practice and thus venerated as far more worthy and politically expressive of civil society than other sorts of practices, especially consuming practices. Green votes are often seen as protest votes, as a way of people voicing their concerns against mainstream parties, rather than support for the non-mainstream parties. Geography plays a large part in this, with protests seemingly more likely in elections that are seen as less important, e.g. at the national and supranational scales, while more positive support and green successes are seen at the local level.

However, many people do not exercise their democratic right to vote, even where it has been hard-won through struggle, whether through not caring, not knowing how to vote, not knowing how to register to vote or even actively refusing to vote as a political statement that rejects the entire system. And amongst those who do bother to vote, most people do not seem to be influenced in their voting by environmental issues, with only a small minority self-identifying as 'green' voters and a very small number belonging to explicitly Green parties. In some countries, green voting is not possible because Green parties do not contest elections and thus do not appear on ballots, and environmental policies are absent, or at least insignificant, in the manifestos of mainstream parties.

It also seems that environmental voting is more obviously separated from other sorts of environmental practice. This happens through time and space: elections are short-lived, geographically highly structured events and the ability to vote, that is, to be entered onto an electoral roll or list held by the state, is also tied strongly to one's residence in a particularly location at a defined time. Compared to practices of enjoying, consuming and working, voting is singular, demarcated and unusual – perhaps even exciting the first time around – rather than a normal, mundane part of everyday life. Unlike other practices that I have covered so far in this book, voting is not repeated very frequently and may be less likely to be routinized through repetition in the way that other practices are, such as driving, recycling, gardening. People are more likely to think consciously and deliberately about how they are going to vote before they do vote, whereas many of the other practices we have looked at often become so routinized that the reasons for them are lost or obscured – they become habits rather than deliberated decisions. That is not to say that people may not still change their minds or make an impulsive decision when they finally find themselves in a polling booth; 'floating voters' or 'protest votes' may still swing an election, although they are usually in the minority.

So I am not arguing that voting is easier, more rational or more predictable than other practices, but I do think voting practices are more explicitly spatialized and less routinized than other practices, meaning that they are more likely to result from people's conscious deliberation, rather than taken-for-granted habits. The effects of geography are also particularly pronounced and traceable for environmental voting, with scalar hierarchies of worth reflected in the results of extensive datasets. However, as with many other practices, the precise reasons for voting are often only articulated post-hoc, if at all, re-emphasising the problems of relying on self-reported data to analyse and understand environmental publics.

Notes

1   In the 2015 contest for the post of leader of the Labour Party, following the Party's poor showing at the General Election in May of that year, the party's membership shot up, more than tripling in only a few weeks, an expression of public interest remarkably out of keeping with recent trends. This was particularly prompted by the fact that becoming a member of the Labour Party meant a person could vote in the leadership contest that year and thus influence who would take over as leader of the party and of opposition to the Conservative Party that was in government.
2   Like other Green Parties, the Greens in Germany do not have a party 'leader', but Fischer was still widely regarded as their leading light in election campaigns in 2002 and 2005.

References

BBC (2014). Greens beat Lib Dems to come forth in European poll. 26 May. www.bbc.co.uk/news/uk-politics-27570514

BBC (2015). Election 2015 results. www.bbc.co.uk/news/election/2015/results

Bomberg, Elizabeth (2001). The US Presidential election: implications for environmental policy. *Environmental Politics* 10, 2, 115–121.

Carter, N., C. Rootes, A. Jordan, J. Fairbrass and D. Toke (2001). Profile 1 – One step forward? Greens and the environment in the 2001 British General Election. *Environmental Politics* 10, 4, 103–120.

Dalton, Russell J. (2005). The greening of the globe? Cross-national levels of environmental group membership. *Environmental Politics* 14, 4, 441–459.

Dalton, Russell J. (2015). Waxing or waning? The changing patterns of environmental activism. *Environmental Politics* 24, 4, 530–552.

Dunlap, Riley E. and Aaron M. McCright (2008). Social movement identity: validating a measure of identification with the environmental movement. *Social Science Quarterly* 89, 5, 1045–1065.

Federal Commission (2013). *Federal Elections 2012: elections results for the U.S. President, the U.S. Senate and the U.S. House of Representatives.* Federal Election Commission, Washington, D.C.

Frankal, Elliot (2005). Compulsory voting around the world. *The Guardian* 4 July. www.theguardian.com/politics/2005/jul/04/voterapathy.uk

Franklin, Mark N. and Wolfgang Rüdig (1995). On the durability of green politics: evidence from the 1989 European Election Study. *Comparative Political Studies* 28, 3, 409–439.

The Green Party (2015). Our people. www.greenparty.org.uk/people.

The Green Party of the United States (2015). All candidates for office. www.gp.org.

Keen, Richard (2015). Briefing paper number SN05125. Membership of UK political parties. (House of Commons Library, London) http://researchbriefings.parliament.uk/ResearchBriefing/Summary/SN05125

Kellner, Peter (2009). Britain's oddest election? *Political Quarterly* 80, 4, 469–478.

Rüdig, Wolfgang (2006). Is government good for Greens? Comparing the electoral effects of government participation in Western and East-Central Europe. *European Journal of Political Research* 45, S127–S154.

Rüdig, Wolfgang (2012). The perennial success of the German Greens. *Environmental Politics* 21, 1, 108–130.

Rüdig, Wolfgang and Mark N. Franklin (1992). Green prospects: the future of Green parties in Britain, France and Germany. 37–58 in Wolfgang Rüdig (edited), Green Politics Two. Edinburgh University Press, Edinburgh.

Rüdig, Wolfgang, Mark N. Franklin and Lynn G. Bennie (1996). Up and down with the Greens: ecology and party politics in Britain 1989–1992. *Electoral Studies* 15, 1, 1–20.

Soper, Kate and Frank Trentman (2008). Introduction. 1–16 in Kate Soper and Frank Trentman (edited), *Citizenship and Consumption.* Palgrave Macmillan, Basingstoke.

Tranter, Bruce (2012). Social and political influences on environmentalism in Australia. *Journal of Sociology* 50, 3, 331–348.

UK Political Info (2015). *General Election* turnout 1945–2015. www.ukpolitical.info/Turnout45.htm

US Census Bureau (2012). *Statistical* abstract of the United States. www.census.gov/prod/2011pubs/12statab/election.pdf

US Census Bureau (2014). Voting and registration. www.census.gov/topics/public-sector/voting.html

US Federal Commission (2013). *Federal Elections 2012. Election Results for the U.S. President, the U.S. Senate and the U.S. House of Representatives.* Washington D.C.

Vasilopoulos, Pavlos and Nicolas Demertzis (2013). The Greek Green voter: environmentalism or protest? *Environmental Politics* 22, 5, 728–738.

8

WORKING PUBLICS

Introduction

My employer is the University of Hull and it has tried to influence how its staff and students work and their environmental impacts for some years now. Its energy manager monitored electricity usage and found that it remained high overnight even when many offices were empty, because many University staff reportedly left their computers running overnight to save time booting them up the next morning. Since then, the University has implemented shut-down programmes on all computers, so that they go into 'sleep' mode when they are not actively used for a period of time. The University has done the same with lighting by installing motion sensors in many corridors and its energy management initiative also encourages staff and students to turn off the lights when leaving rooms. To implement its Carbon Management Plan, the University urges staff to "do your bit by remembering to switch off appliances and rather than taking out the energy-guzzling fan heater, perhaps putting on an extra jumper in winter", and exhorts staff to use videoconferencing to reduce travelling and to travel to work by walking or cycling in the 'Hull UtravelActive' project, instead of by private car (University of Hull 2015).

Like consumers, as we saw in Chapter 4, workers are therefore encouraged to think about the environmental consequences of their practices, but also to implement small changes that reduce environmental impacts and, some might say more importantly, reduce running costs for their employer. But there are differences between consuming in one's own household and consuming while being employed by an organisation: the procedures and social norms may differ, as may the ability of workers to influence decisions. For example, if I want to buy a new computer for work using University funds, I must use an approved supplier and choose from a restricted list of models, whereas if I want to buy a new computer for home using my own money, I am not restricted to the approved list.

This chapter is about these differences and it focusses on environmental publics at work. In many ways, being at work involves the same practices as in the home or leisure time: we travel for work, we eat at work, we operate cars and computers, we buy things at and for work, both for ourselves, such as lunch, but also for the employing organisation by ordering raw material supplies for manufacturing or office supplies for administrative tasks.

So why did I feel it useful to partition out working publics from the consuming publics or enjoying publics that are covered in other chapters? The ways in which decisions are made at work, the control that is perceived and the infrastructure and technologies available all mean that people may develop, share and adopt very different environmental practices when they are at work compared to when they are out shopping, at home cooking or out enjoying themselves. And certainly compared to minority practices like environmental voting, campaigning and participating, far more people are involved as environmental publics through working, second only perhaps to publics involved in consuming in terms of the proportion of the population over the course of their lives.

Practices

When examining working practices, we need to remember that many people feel very differently about being at work compared to being at home, so the meanings of practices shift considerably, as well as the places in which they are performed and the objects and devices incorporated into those practices.

For example, Crowell and Schunn (2014) surveyed local authority employees online and found that people were reportedly far less likely to engage in a range of environmental practices at work than at home, especially practices such as recycling, driving and purchasing. They also found that the level of environmental engagement was more similar when comparing practices in the same context (home or work) than when comparing contexts, that is, there was a stronger correlation between environmental practices at work than there was between one environmental practice, such as recycling, at work and the same practice at home.[1] The authors suggested that "what works to promote environmental conservation actions at home may not work to promote the same at work" (Crowell and Schunn 2014, p. 731), an important consideration for those attempting to shift behaviours for policy reasons.

Another reason to devote a chapter specifically to working publics is that the academic literature is also divided. Researchers in business and organisational studies tend to analyse working practices by focussing on the working organisation, taking "a macro-institutional and/or functionalist perspective that black-boxes the organization" (Costas and Kärreman 2013, p. 411), or by focussing on how to promote 'the greening of industry' through case studies of organisations in journals like *Business Strategy and the Environment* and *Greener Management International*. Such approaches tend to place the organisation centre-stage as the agent of its own practices, acting in and of itself, albeit within the context of other pressures and influences, including regulation. But focussing on 'the organisation' obscures the messy reality of how working practices are co-produced through the collective of myriad, often tiny decisions made by many workers on a daily basis.

I want to shift from grander-scale explanations to look at the more complex, individual and repeated patterns of practices by publics at work. Other social sciences are similarly attempting to open up that 'black box' of the organisation to analyse how the individual people and practices are brought into and develop within the organisation's context, for example, taking approaches from sociology and anthropology. Similarly, economic geographers have been urged to see the "micro-social activities" that make up working practices, so as to make analyses "more sensitive to agency than structuralist or institutionalist accounts of the world" (Jones and Murphy 2011, p. 376).

But even where individual working practices are explicitly analysed, research often focusses upon the top-level managers and strategists, who are assumed to make the important decisions about how a corporation works or to serve as champions of change in their organisations (e.g. Elkington and Burke 1987). Individual entrepreneurs have been cast as innovators of technology and its applications, capable of branding and marketing new trends such as sustainability. An environmental entrepreneur or 'eco-preneur' may experiment with and try to shift practices that are seen as alternative or 'niche' into mainstream operations, disrupting existing practices that have become routinized and entrenched, for example, when choosing the materials used to construct houses in the UK (Gibbs and O'Neill 2014). Holliday et al. (2002, p. 28) referred to entrepreneurs as "creative destroyers", because of their capacity to disrupt and thus change practices. But this perspective becomes problematic when the innovative entrepreneur is conceptualised (perhaps unintentionally) as the lone 'hero' figure, usually male, such as James Dyson or Richard Branson (Gibbs and O'Neill 2014; Prudham 2009).

Rather than these top-level managers and imagined industrial 'heroes', this chapter will concentrate on ordinary people in employment and how their everyday practices become routines, how those routines might be changed and how analysing working publics and practices reflects academic conceptualisations of place and power.

Working as environmental professionals

Let us begin by considering the diversity of working publics and the organisations and roles in which they perform work in relation to the environment. First, some people specifically seek (or end up in) jobs that are environmentally related, such as working for environmental non-government organisations (NGOs) such as the National Trust, WWF or Friends of the Earth. Some staff in environmental NGOs are environmentally focussed workers, e.g. who have environmental qualifications and/or who read environmental journals, commission environmental research, write reports on environmental issues or findings and lobby for environmental reform, but others work in non-environmental roles that keep the organisation going, such as accountants, lawyers, public relations professionals, copy-editors, kitchen staff, cleaners and couriers.

Environmental NGOs are often small and so have few staff. For example, the 139 non-profit environmental organisations in the UK that were surveyed by Cracknell et al. (2013) reported a total of 11,125 FTE employees working on environmental issues (that is, excluding non-environmentally focussed staff like accountants and cleaners), with a median number of only seventeen employees per NGO, although this was skewed by a few very large NGOs who employed most of these environmental workers. Despite this, large organisations tend to dominate our ideas of environmental groups, due to their greater media coverage and public profile. For example, in 2014 the UK's Royal Society for the Protection of Birds (RSPB) reported about 2,217 paid employees on average and another 10,000 unpaid volunteers, and in 2013 Greenpeace International reported 2,830 staff across the world, with the largest contingent (over 1,000) in Europe, making them important environmental employers.

Another major environmental employer is the state, in the form of environmental regulatory agencies and other public bodies. For example, in 2015, the Environment Agency of England and Wales employed about 10,600 people, that is, nearly as many as Cracknell et al.'s (2013) 139 NGOs put together, including both environmental professionals such as geologists, engineers, hydrologists and nuclear assessors, and general professionals such as project managers, planners, policy advisors, enforcement officers, managers and support staff.

There are also many private-sector companies that primarily offer green products and services. Defined as the 'environmental goods and services sector', they include "producers of technologies, goods, and services that measure, control, restore, prevent, treat, minimize, research and sensitise environmental damages to air, water and soil, problems related to waste, noise, biodiversity and landscapes and resource depletion" (Office for National Statistics 2015, p. 23). As well as protection via waste water and waste management, this sector includes environmental education, environmental equipment manufacture, renewable energy, construction, insulation and environmental charities. In 2012, this sector was worth £26.3 billion in the UK and employed 357,200 people (full-time equivalents), with the largest subsector being waste water and waste management services, which employed about 120,600 people (full-time equivalents) (Office for National Statistics 2015, p. 1). Environmental workers in such organisations are represented by professional associations such as the Chartered Institution of Water and Environmental Management (CIWEM) in the UK, which provide specialised training, chartered status and often lobby government on behalf of their members.

And even within companies and state bodies that do not explicitly deliver environmental products and services, there are still staff responsible for environmental management, perhaps as part of their responsibilities for health and safety in the organisation or as part of corporate policy. Large multinational corporations may employ as many as the whole UK environmental goods and services sector put together: in 2015, PricewaterhouseCoopers employed 208,000 people, Cargill 153,000, Procter and Gamble 118,000, Toys 'R' Us 66,000 and Tesco 310,000 in the UK. Only a small minority of these workers will have environmental matters as their primary focus, although all of them contribute to the environmental impact of the organisation as a whole.

In the education sector, universities primarily deliver research and higher education, but many now have environmental committees to develop, implement and monitor their environmental policies and impacts, as we saw in my opening vignette. Every year, universities are ranked on their green reputations by People and Planet (2015) in the UK. One criterion used is that a university should have at least two full-time equivalent members of staff per 5,000 students with at least 50 per cent of their job being responsibility for policy and implementation of environmental management (People and Planet 2015b), again demonstrating that specialised environmental workers are needed even within organisations whose main business is not environmentally focussed.

Working environmentally

Because the environmental roles and organisations that I have mentioned so far are very much in the minority, the majority of working publics do not work in explicitly environment-facing ways, but their working practices still have environmental impacts. Office workers in companies of all sorts make decisions about photocopying, driving to conferences, ordering supplies and so on, decisions which have environmental consequences. In effect, everyone who works is engaging in environmental practices, but often these are unrecognised and unappreciated.

As mentioned, analysing practices has not been a common approach to understanding how working people relate to the environment, with business studies tending to see a corporation as the agent, rather than staff as diverse publics. Research that has studied workers has tended to focus on managers and environmental officers because of their special roles as agents of corporate environmental policy and practice. A good example is work by Stephen Fineman (1996) on the emotions of working individuals and the meanings that environmental behaviour at work has for them. Having interviewed managers in seventeen large UK supermarkets, Fineman concluded that some managers were committed to their company's environmental policy because they wanted to improve their own work, but were also driven by a sense of pride in their professionalism or concern for their company's economic performance, rather than by environmental ethics *per se*. Others championed green causes out of a personal sense of morality, although they often struggled to convince their work colleagues or to change practices in the face of resistance from colleagues.

Some managers in Fineman's study claimed to resist lobbying by environmental NGOs because it was 'blackmail' and defended their own decisions as business decisions driven by customers, that is, they felt that environmental pressures were not valid considerations, echoing Milton Friedman's (1970) famous dictum that the only responsibility of business is to make a profit, not to submit to 'fundamentally subversive' ethical ideas. Similarly, Sarasini and Jacob (2014) interviewed managers in the Swedish electricity industry about their perceptions of climate change and concluded that managers chose to comply with state environmental policy, but largely ignored other external pressures such as customers, civil society or NGO lobbying. To put this another way, managers mainly imagined the future of their company in the context of the expected policy regime, especially its likely costs, rather than in the context of ethical values, and they adapted working practices accordingly.

However, even where managers are not ethically driven, they may feel they have to actively perform in ways that suggest ethics are important, that is, 'talking the talk' of their environmental responsibility as dictated by company policy, rather than feeling it individually. This shows that working publics sometimes enact (and articulate in interviews with researchers) their professional role in their company in line with written corporate policy, even if this means that they must pack away their personal ethics. "Conflict between an actor's private moral belief and corporate green expectations is unusual, and readily resolved in favour of the corporation" (Fineman 1996, p. 496) through their professional attitudes.

And not only do managers perform differently at home and at work, but they also often explicitly recognise these differences, some using the pronoun 'we' not 'I' in research interviews to emphasise that their viewpoint emerges from their socialised, embedded role within their organisation rather than from personal ethics.

Workers therefore internalise organisational cultures of managing and reducing environmental impacts (and costs) through 'good housekeeping' practices of reducing waste, water and energy use. This can be interpreted as a form of internal brainwashing that has more power to change employees than similar appeals have to change consumers or citizens, because employees want to keep their jobs and also (in some cases) be proud of their jobs and their employer. This interpretation supposes that environmental practices are communicated and encouraged in a topdown way (e.g. Brunton et al. 2015) from management to non-management as organisational culture and working norms, leaving ordinary workers little possibility to imagine or implement alternatives, discouraging reflection on and change to existing and normalised routines of working. Inertia is thus often a product of how people perceive their work contexts.

But employers often also encourage their workers to be good citizens, e.g. to raise money for charitable causes and the local community through organising and participating in sponsored events while at work. Sometimes employers 'match-fund' any money raised by employees or give awards to employees to celebrate their charity work or 'donated' time (e.g. Rondinell and Berry 2000). This mixes philanthropy with employee development, local outreach and positive public relations to associate the organisation's name with charitable activities and fund-raising, but it also bundles together practices outside the workplace with those inside, whether this involves workers sponsoring each other to participate in charitable 'fun-runs' or workers collecting donations for a charity from other workers during or after work.

Temminck et al. (2015) analysed such practices by workers in terms of representing 'organizational citizenship behaviour', that is, personal, voluntary practices that go beyond an employee's normal duties and are not explicitly required or rewarded by their employer, practices that might be environmental or social, e.g. supporting other employees. They found that employees in two public-sector organisations were more likely to enact environmental practices at work if they perceived their employing organisation to value their voluntary efforts and if they felt emotionally attached to their employing organisation, again emphasising the importance of work and its meaningfulness in supporting environmental practices.

In this way, workers are encouraged not only to internalise corporate statements of ethics into their working practices, but also to internalise citizenly behaviours within working practices, to a degree at least. This is in addition to the corporation donating to environmental causes directly, which is another common practice claimed as corporate philanthropy.

Changing environmental working

Like domestic practices, there have been efforts to deliberately change working practices to make companies and other employers more environmentally friendly, in the process creating a new subdiscipline and literature about 'the greening of industry'. I want to briefly analyse this notion of change, to show how it is conceptualised.

In the 1970s, environmental concern began to grow and often cast business and industry as the environmental villains, the causes of large-scale environmental disasters such as in Bhopal, India (caused by Union Carbide, now Eveready Industries), Love Canal, USA (caused by the Hooker Chemical Company, see Environmental Protection Agency 2015/1979) and the Exxon Valdez oil spill off Alaska, USA, as well as the cause of much low-level but persistent diffuse pollution from factories, farms and transport. As environmental regulations grew in response, many companies lobbied against them, but since the late 1980s, some business people and commentators began to react differently, encouraging business to take their responsibility to the environment more seriously.

The chemical industry's Responsible Care initiative, spread through national sector associations, was one response to the industry's poor reputation, one that sought to instil better practice through sharing information and pledging good conduct and helped companies to anticipate and manage risks more effectively. John Elkington also wrote extensively about the potential for the greening of consumers and business, while running his own environmental communications consultancy. For example, in 1987 he co-wrote *The Green Capitalists* with Tom Burke, then Director of The Green Alliance, a newish environmental NGO in the UK, and later an advisor for government ministers, corporate interests and universities. The book's mix of environmentalism and business advice proved immensely popular, and helped to set an agenda for business research, advocacy and practical advice that continues to this day. The journal *Business Strategy and the Environment* was launched in 1992 and was followed by others as the literature expanded.

Implicitly, such writings and advocacy attempted to change industry from the environmental villain to a sort of environmental saviour, through changing its approach to environmental management. Advocates argued that organisations did not need to surrender self-interest to altruism to be 'greener' because they could also benefit from such changes. There have also been lots of 'how to do it' reports over the decades, often using case studies from well-known organisations.

For example, in 2002, the CEO of Dupont, the Chairman of Royal Dutch/ Shell and the founder of the Business Council for Sustainable Development published *Walking the Talk* (Holliday et al. 2002) as a strategic input to the UN World Summit for Sustainable Development in the same year. Reading their book, we see business leaders arguing for the globalisation of markets as the best way to achieve sustainability, as well as advocating corporate social responsibility not for its own sake, but for the sake of competition and profit (Holliday et al. 2002, p. 108). Citing not academic theory but practical case studies of what companies do or do not do, their approach focussed on the leaders of corporations, quoting them rather than ordinary workers and their everyday practices. They wrote about corporations and their workers being good 'citizens' through improving environmental management. Similarly, Elkington and Burke (1987, p. 233) quoted Sir Peter Parker, then Chairman of Rockware and the British Institute of Management saying that "people in our companies are not just employees, not just trade unionists … they are citizens at work, so the message of environmental quality must matter to them. At work, we are all environmentalists now".

But the message of environmental quality does not necessarily matter to all employees and not all workers are environmentalists, neither back in 1987 nor today. Claiming industry to be the saviour of the environment is as flawed as expecting consumers to be its saviour, or voters, or any other diverse set of actors. Understanding working publics means going beyond organisational strategies and mission statements to look at the far more messy and fragmented practices of people and things within organisations. Changing multitudinous, often routinized or even unarticulated and forgotten practices cannot be taught in a one-hour webinar, nor changed across a corporation of thousands of employees through newsletters and corporate vision statements. Instead, working practices are configured, distributed and (in some cases) lost in interactions between diverse people and things that spiral round and about over time.

Changing working practices is therefore difficult, because established norms need to be disrupted and re-learned. For example, Holland et al. (2006) found that being asked to plan how to implement recycling at work could break office workers' routinized practices and implement new ones far more effectively than merely being given information and encouragement. Here, the deliberate intervention that asked workers to plan out new ways of working for themselves, explicitly and with specific goals for recycling, disrupted the routinized practices of *not* recycling and caused workers to design and gradually routinize new recycling practices.

Individual workers may also disrupt 'normal' practices, intentionally or not, and thus expose the possibilities for change. In a study of low carbon working, a manager in a cold-storage company was interviewed by my colleague and he told a story of switching off what he thought was an unnecessary light before going home, to reduce its high heat output and consequent environmental impact. But because these lights were 'normally' on permanently, 24 hours a day, seven days a week, the night workers assumed that there had been a power failure and called out an engineer at midnight to fix them. Although slightly embarrassing for the manager, this also resulted in the company installing movement sensors on lights to avoid them being on permanently, and changing many lights to a lower power type.[2]

Routinizing any practices also often depends on things and infrastructure (as we saw in Chapter 1), such as independent heating and lighting controls or the provision and proximity of bins for rubbish and for recycling, as well as cultural expectations and norms about which practices are appropriate or not in the workplace. In the university where I work, recycling bins were put on office corridors at the same time as cleaning staff stopped emptying the bins in individual offices, so that not only were things rearranged, but the practices of filling and emptying them shifted meaning by moving responsibility from cleaning staff to office staff.

Practices are de-normalised not solely by the example of individual champions or by providing information, but by physically disrupting and reconfiguring spaces through which practices occur, to make some practices physically easier or harder for some practitioners. In some cases, de-normalisation is resisted and normal practices re-asserted, as in the failure of environmental champions to initiate a 'No Bin Day' in the workplace studied by Hargreaves (2011). Hence, any 'greening of industry' is not merely a greening of corporate policies, but a greening of working publics and their everyday infrastructure – a much more complex process.

Codifying environmental practices at work

Despite this, there have been attempts to implement a normative agenda of encouraging environmental management and responsibility through corporate policies and practices. Various forms of environmental management system (EMS) exist that specify practices for organisations to follow, so that, rather than developing their own practices or learning from their friends and family (as households tend to do), organisations can follow a manual of standardised practices to monitor, evaluate and ameliorate their environmental impacts. In contrast with environmental practices of consuming, for instance, environmental of working tend to be codified, written down and even specified in terms of measurable practices.

One example is the ISO 14000 family of standards for environmental management, which has been adopted by many organisations across the world (see ISO 2015). Achieving these standards requires organisations to:

- define their objectives, i.e. practices of meeting, discussing, writing and communicating;
- review existing operations, i.e. practices of gathering data on environmental emissions and risks, writing reports, meeting, discussing information and gaps;
- ensure support from the organisation's leaders, i.e. communicating commitment;
- communicate environmental practices internally to the workforce and externally to shareholders, regulators and the wider public (in person, in writing, on paper and online).

For example, Sony achieved ISO 15001 for all its manufacturing sites by 2000, as part of its environmental initiative, Green Management 2005 (Holliday et al. 2002, p. 147). Organisations can choose to subcontract (and pay) third parties, such as accounting firms or NGOs, to certify that they meet such standards, although this is not compulsory; many organisations self-certify to avoid the fees incurred by subcontracting. Howsoever it is accomplished, achieving ISO14001 can be used as a 'reputational asset' (Perkins and Neumayer 2010, p. 339) in corporate PR.

Standardised environmental management practices can give somewhat fuzzy notions like sustainability a more concrete and coherent form (Lahneman 2015, p. 177). Codifying standards in terms of practices that can be directly implemented enables staff to share these practices more easily: (a) within an organisation, even where that organisation operates in multiple nation-states with different regimes and cultures, (b) across organisations but within an industrial sector and (c) between an organisation and its suppliers (and would-be suppliers). Such practices should persist even when staff working for the organisation leave or move jobs internally or when new staff join, because certification should confirm that the new practices are becoming normalised, for example, through training courses for existing staff or induction events for new staff.

Yet in practice, different staff are going to perform these practices differently, so codification of practices is often incomplete or not applied in practice. For example, the Forest Stewardship Council was set up to promote sustainable forestry in the early 1990s through setting up certification standards. These were written by their staff, certified by a range of companies and environmental NGOs and covered all aspects of the forest supply chain, from forest management and logging to pulping, paper-making and book printing. FSC's (2000) principles and criteria were intended to be general enough to be universally applicable to any region or organisation across the world, but the standards against which organisations would be assessed needed to be more measurable and adaptable to particular places, requiring some translation of the principles into practices that could be interpreted and implemented in specific contexts.

Interpreting these written FSC standards into organisational practice is performed by individual workers, not by the organisation *per se*. If a forest manager wants to have their forest certified as meeting the FSC standard, they invite (and pay) an independent auditor to review their paperwork and visit their site to see how they manage trees, and this may involve negotiation in the field and in the office between the forester and the auditor. In one forest in Scotland (see Eden 2008), I observed an auditor discussing with a forester how many trees should be left standing after clear-cut logging, to provide the 'deadwood' required by the FSC standard to support populations of insects and birds and thus to sustain biodiversity. The forester told the auditor that the logging contractor had been asked to leave two or three trees of standing deadwood per hectare, because this is what the FSC standard stated, or, if none were standing dead, to cut some live trees off at 6 feet high and leave them to die standing (these are called 'snags'). The auditor said that it would better for biodiversity to leave "a clump" of standing deadwood, rather than three separate trees per hectare, thus adapting the meaning and practice of the standard in a particular geographical context. Similarly, Callon et al. (2002) argued that markets and socioeconomic practices are made (often implicitly) through accounting practices and metrologies of calculation, measurement and commensurability. FSC's standards thus aim to challenge and disrupt working practices, in order to shift them towards more sustainable ones, but exactly how this is done depends upon the mutual performance of the workers involved.

As well as negotiating practices, sometimes it is the job of specific workers to persuade other workers to enact environmental practices in line with the organisation's policy, e.g. switch off lights, minimise travel for business meetings, follow risk assessment procedures and so on, even to the point of monitoring other employees to check that they comply with the specified norms. Hargreaves (2011, p. 86) analysed the volunteer 'Environmental Champions' in a large UK company and found that they recognised their own surveillance role and found it uncomfortable, because they had to highlight, challenge and disrupt 'normal' practices. They also performed as exemplars, by demonstrating the new, 'better' practices to their colleagues through their own working, to try to jolt others into copying and thus sharing those practices. Hence, some working publics also explicitly set out to shape other working publics within their own organisation or, sometimes, through a business association or other business-facing environmental initiative. Such learning from the example of others has also been seen in households and neighbourhoods, but one important difference between those domestic practices and the practices of working publics is that working practices are more likely to be written down in codes of conduct within the organisation, and thus formalised to a degree unlikely in more informal, domestic settings. That is not to say that the written codes are always followed, as I shall discuss later.

Place

Let us move on now to consider the geographies of the diverse environmental practices enacted by working publics. First, working publics have geographies in terms of where their workplace is located. Patterns of employment obviously vary by country and region, but also by whether a workplace is in the city, e.g. financial and retail workers, the countryside with agricultural and tourism workers; or the suburbs, e.g. people who work from home or in other people's homes as cleaners, childminders and gardeners. Commuting to and from work similarly involves moving between these different spaces – some people fly to work in another country, some people take a train to work in a city, some people work from home – with the concomitant environmental impacts.

And the geographical context itself shapes these patterns. For example, multinational corporations may re-locate to another country to exploit not only the geographical differences in natural resource availability and labour costs, but also the differences in regulatory regimes, perhaps moving to a country that has less regulation or lower standards for emissions than the home country. This is expressed in the concept of 'pollution havens', a spatial fix to capital's problems of the rising costs for controlling environmental pollution. Significant risk in terms of liability for monetary compensation can also be an issue for companies operating in wealthier countries. Under the USA's Clean Water Act, the Deepwater Horizon oil spill in the Gulf of Mexico has so far cost BP over \$18 billion[3] in fines and compensation. Moving to a less regulated (and usually less wealthy) country can save companies money and allow them to emit more pollution.

Geographical mobility of workers also means that working practices can travel globally, involving assemblages of personnel, technologies and expectations or cultural norms. These geographies can be mapped indirectly using proxy measures, such as whether organisations have been certified to an environmental standard. For example, Perkins and Neumayer (2010) found that the number of ISO 14000 standards certified was higher in countries with more international exports, more international business travel and more foreign direct investment, and that this correlation was stronger in wealthier countries (measured by GDP per head, another proxy measure) in the global North. By comparison, they found that the number of signatories to the UN Global Compact code of conduct, a pledge to ethical standards, was also higher in countries with more exports and international travel, but also in countries with higher scores on a democracy index, rather than on measures of wealth. So different global norms or pledges can be taken up more in some countries than others, producing international patterns of varying compliance.

As well as showing geographical patterns, this suggests that practices of seeking and achieving environmental standards spread geographically not through some national-level edict or advocacy by a business association, but directly through people in one country networking with those in another country, either in person through international business travel or through emails, phone calls, supply contracts, conferences and other ways of working with others. To put this another way, changing environmental working practices is enabled by other working practices, with more mobile and networked working publics being more able to spread and share practices.

In addition, multinational corporations themselves may have internal geographies that share practices across international borders. Many supply chains are internal to large manufacturing groups, so that 'chains of custody' for certification can be managed within the same organisation globally, with practices written down and audited in the same way by those working for the same organisation but in different nation-states. In such cases, working publics learn from others working within the same organisation but in other countries, so that practices are shared through organisational rather than national scales of reference.

A second, quite different geography is one that I have already briefly mentioned: how working practices differ by domains of practice, e.g. the workplace compared with the home or other casual setting such as an internet café. Home and work are both private spaces in the sense that they are closed to others outside of the family or work grouping, but they are also shared spaces where people adopt, adapt and develop practices collectively and routinize these through repetition. In working spaces, decisions over what to buy, how to use what is bought and how to dispose of it may be deliberated through group discussions or enacted without conscious deliberation through habits learned over time or rules written down for all employees to enact. And people often behave differently when doing the same practice in the workplace compared with the home, as we saw earlier (and Crowell and Schunn 2014).

Another more liminal space is business travelling, a time when people are working, using laptops or phones on trains and airplanes, for instance, or attending conferences elsewhere to be trained in working skills or to hear about and share environmental practices. The mobility of working publics encompasses private and public spaces, diversifying the domains in which working practices are enacted and travel through space. Such practices also enact those spaces: airports, trains, conference suites and cafés today often provide free Wi-Fi to customers, making those spaces more conducive to off-site working and in turn peopling those spaces with more working publics. And these choices made about how to travel to, from and during work also directly influence the environmental impact of working publics, such as whether to fly, drive, walk or take a taxi (e.g. Figure 8.1, taken in Edinburgh while I was travelling as part of my job).

FIGURE 8.1 Carbon neutral notice in a taxi cab in Edinburgh.

We might also contrast working domains by type of employer, as this context can also influence which practices are enacted and how. In the UK, the public sector is a major employer, with about 1.5 million people employed in the National Health Service, nearly 1.6 million in state education and over 1 million in other 'public administration' jobs in 2015 (Economic and Social Research Council 2015). Research into personal environmental attitudes has often suggested that those working in caring professions such as healthcare and education express more environmental attitudes and act in ways that are more pro-environmental. But the business studies literature has tended to analyse public-sector and private-sector organisations in similar ways, reasoning that a large university and a large retail corporation, for instance, show a lot of similarity in practices of procurement, operations, human resources, accounting and so on, despite having quite different aims in serving their learners or customers. Again, it is the context of working that is argued to differentiate practices from those at home or at play, rather than the type of organisation for which one works.

A third way in which geography shapes working practices is through shifts to online practices, so that workers collaborate through the cyberspace of the internet, instead of moving physically through space. In what we could call a 'digital fix' to the spatial problem of the huge environmental impacts of business travel, many organisations (including my own employer) now encourage staff to shift from flying and driving to telecommuting and e-conferencing, to reduce costs but also environmental impacts. As well as reducing travelling or changing its geographical patterns, this also means that organisations purchase, set up and operate new in-house infrastructures in the form of video-conferencing suites with webcams and large screens, and may upgrade computing facilities to support secure dial-in working from home. And working publics may also reconfigure their own domestic space to create a home office with computer, printer, phone and desktop, as well as changing travelling times, patterns and sociality, thus consuming while also producing – another example of 'prosumption', the blurring of consumption and production that we met in Chapter 4. The presence of the employee at work can thus be digital rather than embodied, changing not only the physical infrastructure but also the social and psychological character of work through differently embodied or mediated interactions between working publics.

Finally, geographies of working are also shaped by constructions of scale. As in other chapters, geographies of working practices often invoke a scalar hierarchy as an organising principle, differentially weighting some scales as more important, powerful or self-evident than others. In the case of working publics, a typical scalar hierarchy underlying analysis is:

- the national scale of the state or regulatory regime as most influential;
- the scale of the organisation and its cultural norms and routines;
- the scale of the department or working team within the organisation (particularly whether there is an 'environmental' department/team or not);
- the micro-scale of the individual worker as least influential.

One problem with this particular scalar hierarchy is that the scale of the organisation is highly variable, from micro-businesses with five or fewer employees to multinational corporations or state agencies with thousands of employees spread across several countries. Yet this is often the scale that is analysed in business studies, conceptualising (whether explicitly or not) the organisation as a quasi-entity, an actor with attributed identity, agency and strategy. This delusion is surprisingly easy to sustain in academic and non-academic writing, despite it being quite obvious on a moment's reflection that it is not the organisation but working people within the organisation that *do* things. In this chapter, I have instead shown how working publics and their practices are mutually performative, both formally through corporate codes of conduct that are written down and management systems of sanctions or incentives, and more informally through organisational norms and teams sharing practices when collaborating to complete tasks. I have therefore emphasised not the large strategic decisions, but the many smaller decisions made by people while in the work space, whether their presence is physical or virtual.

Power

Finally, let us consider the implications of the discussion so far for conceptualising the power relations of working publics. I have already suggested that one of the ways that working practices differ from the consuming, knowing or participating practices covered elsewhere in this book is that working practices are performed within the organisational context, while the individual is paid by the organisation. This can restrict performativity if workers perceive that the social norms in a workplace prevent them instigating change; instead, control is often perceived to lie primarily with senior managers through codified systems of purchasing and travelling, as well as more everyday practicalities such as how to heat, cool and light the workplace. Explicit corporate procedures and standards for procurement, travel and risk assessment are all written down and used deliberately to shape how working publics enact their daily routines, in some cases for good reasons to do with avoiding accidents and protecting the workforce, but with the side-effect of restricting diversity and innovation in practices, such as choosing a less environmentally damaging or more energy-efficient product from a source which is not on the employer's 'approved' list of suppliers.

Researchers have used political economy approaches to argue that that this control of workers serves the interests of capital in the form of business owners and shareholders, and that the 'greening of industry' is therefore not part of a gradual transition to a more environmentally sustainable future, but merely a public relations exercise in 'greenwashing' (e.g. Athanasiou 1996). Such approaches suggest that environmental rhetoric is used to obscure or disguise attempts at capitalist accumulation and the pursuit of greater profits through the exploitation of workers and the environment. For example, McCulloch (1990, p. 218) argued that

> *The Green Capitalists* [referred to earlier] will appeal to those who want a concrete example of the way in which the system incorporates, makes harmless,

and then makes a profit out of movements and ideas which seek to challenge it by selling the idea back to the activity in the form of one of more commodities.

As well as seeking to control individual workers, such approaches also see businesses as seeking to control discourses and ways of thinking about environmental protection that have proved powerful in shaping agendas in policy and practice. This "hi-jacking" (Welford 1997) of environmental issues for the benefit of business – what Beder (1997, p. 234) called "the corporate subversion of the green movement" – can also involve business-led and business-financed lobbying of governments in order to avoid, weaken or repeal environmental regulation of business activities.

At a global scale, Sklair (2001, p. 207) argued that the "transnational capitalist class" sought to appropriate the concept of sustainable development and redefine it to suit the interests of global capitalism, especially by offering solutions to it that encouraged more consumption. He defined the "transnational environmentalist elite" as the key agents of this power, through an alliance of four groups: "transnational environmental organization executives and their local affiliates, globo-localizing green bureaucrats, green politicians and professionals, and green media and merchants" (Sklair 2001, p. 210), linking the likes of WWF, the UN and Bill Clinton with top executives from companies like Shell, BP, DuPont, B&Q and Sainsbury's supermarkets. Whilst persuasive, Sklair's (2001) top-down analysis assumes that the practices of working publics and the meanings given to these are determined by senior management through global collusion, rather than by the routinisation of millions of mundane decisions made by working publics within organisations every day.

Researchers have also deployed concepts such as governmentality to theorise how consumers are coerced into self-regulating themselves as subjects of the dominant sustainability discourse promulgated by the state and corporations and how, by becoming 'good citizens', consumers relieve the state of responsibility for environmental reform while failing to challenge the power of the existing and oppressive capitalist system (e.g. Rutherford 2007; Paterson and Stripple 2010). McCarthy and Prudham (2004, p. 276) argued that "the hegemony of neoliberalism is made most evident by the ways in which profoundly political and ideological projects have successfully masqueraded as a set of objective, natural, and technocratic truisms". To put this another way, they are arguing that advocates of corporate environmentalism have tried to normalise the idea that continued economic growth is good for all and to present markets as the natural, sensible, even inevitable solution to environmentally bad working practices, rather than as constructions that need to be challenged and transformed.

In such critiques, the workers or employees of companies (especially large corporations) are conceived as lacking agency, alienated from the products of their own labour; lacking pride in or failing to gain esteem from their lowly functions on an assembly line or office, they are seen to be controlled by corporate managers not merely in terms of tasks set, but also in terms of whether they can join a union, what time breaks they are allowed, the poor conditions at work (pollution, risks taken) they have to endure and of course income that they can earn.

Such analyses thus interpret the corporation as shaping its workers to suit its own ends, making them subject to its rules and desires. This may be done through written rules or through 'soft power', that is, the unwritten norms of acceptable behaviour that develop within each organisation and push workers to fall into line with corporate policy and practices (Brunton et al. 2015, p. 3). So 'greening' an organisation may involve mobilising employees to enact environmental practices set by corporate strategists, implicitly conditioning them to respond in certain ways in line with organisational cultures.

For example, programmes for corporate social responsibility (CSR) are argued to control and regulate workers through managing meaning and identity and such 'aspirational control' (Costas and Kärreman 2013, p. 395) encourages workers to identify with the ideal of a good corporate citizen, to acquire the skills necessary to fulfil that ideal, and to pursue rewards for compliance with that ideal that the organisation offers, in the form of promotions or salary bonuses. Like in Chapter 4, this uses the binary of consumer versus citizen to distinguish between practices, constructing the corporate citizen as more worthy, more democratic and less self-interested, even when employed within a for-profit company. This sort of social engineering seeks to promote compliance with environmentally oriented organisational norms using soft power, rather than explicit coercion.

> Aspirational control typically subtly disciplines and normalizes through attribution, classification and ranking mechanisms … It shifts agency to the target of control, thus converting the individual into an accomplice through confession-like mechanisms, such as self-evaluations and career coaching.
>
> *Costas and Kärreman 2013, p. 398*

Professional development, appraisals and mentoring can all be used as forms of surveillance and self-disciplining (Costas and Kärreman 2013), deploying soft power to internalise corporate exhortations to change. In this way, 'green governmentality' can be used to create compliant environmental subjects (e.g. Rutherford 2007; Paterson and Stripple 2010). Supporting this are the myriad calculative and other metrological practices of "mapping, measuring, organizing, quantifying and above all representing particular aspects of nature" that render the environment an object of analysis and governance (Rutherford 2007, p. 297; e.g. for forestry, Demeritt 2001, or human populations, Hannah 2001). For example, measuring environmental impact through corporate environmental management systems and audits, as discussed above, is often claimed to be the first step towards successfully managing that environmental impact.

And business also shapes environmental agendas and influence consumers through advertising, producing educational material for schools and influencing news stories in the mass media, all of which seek to normalise consumerism in contemporary society. Workers in product design, packaging, marketing and advertising thus shape consuming publics through offering multiple identities for consumers to adopt and adapt by buying and using products and services, ranging from clothing to plastic surgery.

But lest we get carried away with the power of corporate greenwashing, we need to remember what we saw in Chapter 4, that is, many consumers do not do what they are encouraged to do by the state or by corporate advertising; rather, they resist the supposed soft control of governmentality. Many workers likewise resist corporate controls, even while in the workplace, from exploiting the perks of hospitality budgets or making exaggerated expenses claims to not switching lights and computers off to save energy, cost and environmental emissions. And resistance to environmental practices (or any practices) that are urged on the workforce by corporate strategists can be prompted and sustained by apathy, disillusionment and/or disempowerment felt by workers unsatisfied by their jobs, feeling ignored or belittled by managers or passed over for promotion. The control of working by green governmentality is thus incomplete, which is why it continues to be debated and revised (e.g. Rutherford 2007, p. 300).

Summing up

This chapter has shown how working publics relate to the environment through practices that are in many ways similar to those in other spheres of life, but the meanings and power implications often differ because of the context of work. Rather than see 'the organisation' as the practitioner and thus the agent of change, as both business studies literature and political economy approaches tend to do, this chapter has put how ordinary workers at the centre of analysis, to examine how their practices are shared through organisational cultures, travel across space and sectoral domains and are sometimes resisted.

Practices are given meaning and employees given identity through organisational cultures, which may have routinized practices that are habitually performed, rather than deliberately chosen. These shared cultures and practices can be difficult to change, sometimes because employees have (or perceive themselves to have) little agency to change entrenched or codified practices of buying, operating, disposing and so on, or because strategic interventions by top-level management are not translated into new everyday routines. This provokes commentators to argue that for-profit organisations in particular 'talk the talk' of environmental change, but do not 'walk the walk', that is, in words and written policies they may change, but in practices they do not.

Working publics are thus subject to different forms of constraints and control from publics in other sorts of contexts. Some evidence also suggests that workers more consciously adopt particular roles and practices as part of their job, speaking or doing on behalf of the organisation rather than on behalf of their own personal ethics. This has also been encouraged by corporate practices of environmental management and social responsibility, although these do not preclude the possibility of resistance by working publics.

Notes

1   "For example, the home vs. work correlation for recycling is $r = .23$, but the correlations of recycling at home with other home action frequencies is $r = .27$ and recycling at work with other work action frequencies is $r = .39$" (Crowell and Schunn 2014, p. 275).

2 Thanks to Paul Barratt for this anecdote from an interview conducted in 2010 as part of the Low Carbon Futures project.
3 "$5.5 billion under the Clean Water Act (CWA), $7.1 billion to the federal government and the five Gulf states for damages to natural resources, $4.9 billion to settle economic and other claims made by the five states, and up to $1 billion to 400 local government bodies" (*The Economist* 2015).

References

Athanasiou, Tom (1996). The age of greenwashing. *Capitalism Nature Socialism* 7, 1, 1–36.

Beder, Sharon (1997). *Global Spin: the corporate assault on environmentalism*. Green Books, Totnes.

Brunton, Margaret, Gabriel Eweje and Nazim Taskin (2015). Communicating corporate social responsibility to internal stakeholders: walking the walk or just talking the talk? *business strategy and the environment*. Online only: http://onlinelibrary.wiley.com/doi/10.1002/bse.1889/full

Callon, Michel, Cécile Méadel and Vololona Rabeharisoa (2002). The economy of qualities. *Economy and Society* 31, 2, 194–217.

Costas, Jana and Dan Kärreman (2013). Conscience as control: managing employees through CSR. *Organization* 20, 3, 394–415.

Cracknell, Jon, Florence Miller, and Harriet Williams (2013). Passionate collaboration? Taking the pulse of the UK environmental sector. The Environmental Funders Network. www.greenfunders.org/wp-content/uploads/Passionate-Collaboration-Full-Report.pdf

Crowell, Amanda and Christian Schunn (2014). Scientifically literate action: Key barriers and facilitators across context and content. *Public Understanding of Science* 23, 718–733.

Demeritt, David (2001). Scientific forest conservation and the statistical picturing of nature's limits in the Progressive-era United States. *Environment and Planning D: Society and Space* 19, 431–459.

The Economist (2015). A costly mistake. 2 July. www.economist.com/news/business-and-finance/21656847-costly-mistake

Economic and Social Research Council (2015). The UK by numbers: employment. *Society Now* 23, Autumn, 20–21.

Eden, Sally (2008). Being fieldworthy: environmental knowledge practices and the space of the field in forest certification. *Environment & Planning D: Society & Space* 26, 1018–1035.

Elkington, John and Tom Burke (1987). *The Green Capitalists*. Victor Gollancz, London.

Environment Agency of England and Wales (2015). Working for EA. www.gov.uk/government/organisations/environment-agency/about/recruitment

Environmental Protection Agency (2015/1979 originally). The Love Canal tragedy. www2.epa.gov/aboutepa/love-canal-tragedy

Fineman, Stephen (1996). Emotional subtexts in corporate greening. *Organization Studies* 17, 3, 479–500.

Friedman, Milton (1970). The social responsibility of business is to increase its profits. *The New York Times Magazine,* 13 September.

Gibbs, David and Kirstie O'Neill (2014). Rethinking sociotechnical transitions and green entrepreneurship: the potential for transformative change in the green building sector. *Environment and Planning A* 46, 1088–1107.

Hannah, Matthew G (2001). Sampling and the politics of representation in US Census 2000. *Environment and Planning D: Society and Space* 19, 515–534.

Hargreaves, Tom (2011). Practice-ing behaviour change: applying social practice theory to pro-environmental behaviour change. *Journal of Consumer Culture* 11, 79–99.

Holland, Rob W., Henk Aarts and Daan Langendam (2006). Breaking and creating habits on the working floor: a field-experiment on the power of implementation intentions. *Journal of Experimental Social Psychology* 42, 776–783.

Holliday, Charles O., Stephan Schmidheiny and Philip Watts (2002). *Walking the Talk: the business case for sustainable development.* Greenleaf Publishing, Sheffield.

ISO (2015). ISO 14000 – environmental management. www.iso.org/iso/home/standards/management-standards/iso14000.htm

Jones, A. and Murphy, J.T. (2011). Theorizing practice in economic geography: foundations, challenges and possibilities. *Program in Human Geography* 35, 3, 366–392.

Lahneman, Brooke (2015). In vino veritas: understanding sustainability with Environmental Certified Management Standards. *Organization & Environment* 28, 2, 160–180.

McCarthy, James and Scott Prudham (2004). Neoliberal nature and the nature of neoliberalism. *Geoforum* 35, 275–283.

McCulloch, Alistair (1990). Mirror, mirror on the wall: who's the greenest of us all? 210–218 in Wolfgang Rüdig (edited), *Green Politics One.* Edinburgh University Press, Edinburgh.

Office for National Statistics (2015). UK Environmental Goods and Services Sector (EGSS): 2010–2012.OfficeforNationalStatistics,London.www.ons.gov.uk/ons/rel/environmental/uk-environmental-accounts/goods-and-services-sector--egss---2010-2012/uk-environmental-goods-and-services-sector--egss---2010-2012.html

Paterson, Matthew and Johannes Stripple (2010). My Space: governing individuals' carbon emissions. *Environment and Planning D: Society and Space* 28, 341–362.

People and Planet (2015). University league: The 2014 guide. https://peopleandplanet.org/dl/greenleagueguide2014.pdf

Perkins, Richard and Eric Neumayer (2010). Geographic variations in the early diffusion of corporate voluntary standards: comparing ISO14001 and the Global Compact. *Environment and Planning A* 42, 347–365.

Prudham, Scott (2009). Pimping climate change: Richard Branson, global warming, the performance of green capitalism. *Environment and Planning A* 41, 1594–1613.

Rondinelli, Dennis A. and Michael A. Berry (2000). Environmental citizenship in multinational corporations: social responsibility and sustainable development. *European Management Journal* 18, 1, 70–84.

Rutherford, Stephanie (2007). Green governmentality: insights and opportunities in the study of nature's rule. *Progress in Human Geography* 31, 3, 291–307.

Sarasini, Steven and Merle Jacob (2014). Past, present, or future? Managers' temporal orientations and corporate climate action in the Swedish electricity sector. *Organization & Environment* 27, 3, 242–262.

Sklair, Leslie (2001). *The Transnational Capitalist Class.* Blackwell, Oxford.

Temminck, Elisha, Kathryn Mearns and Laura Fruhen (2015). Motivating employees towards sustainable behaviour. *Business Strategy and the Environment* 24, 402–412.

Tesco (2015). Key facts. www.tescoplc.com/index.asp?pageid=71

University of Hull (2015). Carbon management and water. www2.hull.ac.uk/administration/sustainabilityhub/sustainability/carbonmanagementandwater.aspx

Welford, Richard (1997). *Hijacking Environmentalism: corporate responses to sustainable development.* Routledge, London.

9
CONCLUSIONS

Introduction

In most books, the final chapter provides conclusions, summing up the arguments and perhaps speculating beyond the evidence and into the future. Here, I will do some of that, but I also want to mix up the previous chapters, to emphasise that the neat divisions I have used to produce digestible sections of text are themselves part of the problem of getting to know environmental publics more fully. So in this final chapter, I want to consider the ways in which practices are linked between different contexts in which publics work, live and play, including online and offline spaces.

I particularly want to emphasise that geographies of practice are important but neglected in much research into environmental publics and practices, which is why I have explicitly concentrated on the geographies of how environmental publics are made and make themselves in previous chapters. But as well as considering how environmental publics differ through and in *space*, in this final Chapter I also want to consider the vexed question of how environmental publics and their practices change over time, that is, how new practices might be learned, spread and promoted.

Linking practices

Let us start with the linkages between contexts in which publics practise environmental thinking and knowing, being and doing, contexts which I have separated by chapter. For example, does studying an environmental subject during formal education (Chapter 2) make a person more likely to vote green in elections (Chapter 7) or participate in public engagement exercises run by their local authority (Chapter 3)? Does doing outdoor sports (Chapter 5) make a person more knowledgeable about the environment (Chapter 2) and/or more likely to campaign against threats to it (Chapter 6)? Does doing more recycling at home (Chapter 4) make a person think more about and try to increase recycling in their workplace (Chapter 8)? Conversely, does implementing environmental purchasing at work (Chapter 8) make a person more likely to buy recycled paper to use at home (Chapter 4)?

Some research has attempted to answer these questions by linking domains of practice across time and space, e.g. demonstrating that people at work are less likely to choose environmentally related actions than people at home. But research has been hampered by a reliance on self-reporting in surveys, which limits the quality of results and cannot easily track all the routine decisions made by a person that have environmental impacts, nor unarticulated actions and meanings.

Also, different domains are often addressed by different academic (sub)disciplines and literatures, each with their own conferences and networks (see Table 9.1), making it difficult to join them up into an integrated picture of environmental publics across multiple domains.

This is made worse when disciplinary approaches fall into the trap of personifying environmental practices in the idealised form of a 'green voter' or 'green consumer', an imagined figure that implicitly shapes policy approaches and practical initiatives to change how people vote, consume, work and so on. Imaginaries of this sort tend to focus on one domain and one associated set of practices, rather than considering how the context of practices connects with others. In this book, I have tried to unpack such idealised characters by emphasising the diversity within domains, exemplified in my use of the plural term 'publics' in the title and content of this book, as well as my focus on what different people do, say and think.

TABLE 9.1 Domains of environmental practices addressed by different disciplines

Domains of environmental practices	Disciplinary or subdisciplinary literatures that study each domain	Bridging literatures
Home	sociology, anthropology, human geography	
Work	business studies, management, (organisational) sociology	
Play	sociology, business studies, human geography, psychology	tourism and leisure studies
The mind	(environmental) psychology, (environmental) sociology	attitude–behaviour gap
Shopping	sociology, human geography, anthropology	material culture
Travelling	sociology, business studies, logistics	mobilities
Public realm of voting, campaigning etc.	political science, sociology	
Online	sociology, computer science	human–computer interaction

That said, focussing on practices rather than people can throw up its own problems. Studies using social practice theory can feel rather depersonalised, missing the vitality and negotiation between people routinely involved in the same practices, such as how the members of a household who have lived and eaten together for years still argue over what foodstuffs to buy when out shopping. Indeed, almost as soon as researchers choose to focus on one aspect of environmental consequence in how people live their lives, other aspects begin to be obscured, emphasising the complexity involved.

Geographies of environmental publics

As I have already mentioned, I have sought to emphasise the geographies of practice in previous chapters, to address a gap in much of the recent research into environmental publics and practices. While there is some research into geographical aspects of these, such as differences between countries in terms of how their publics think about environmental issues and act environmentally, the spatialisation involved is often implicit and rarely addressed conceptually, although there are exceptions (e.g. Devine-Wright 2013).

In particular, I have shown how many interpretations of environmental publics draw on implicit scalar hierarchies of worth, with important implications for the power of those publics to influence others as well as to influence policies. Usually these hierarchies put national and international scales of activity and impact far above those of the local and regional, denigrating the latter as a consequence and neglecting to analyse how practices are routinized, disrupted and shared precisely through the many mundane, small-scale decisions and undeliberated choices of environmental publics.

Appreciating the spatialisation of practices also takes us beyond international differences to consider also different spaces of practices, the contexts and settings that shape how meanings are developed and attributed to different practices by different publics. Spaces of home, work and play are made through the performance of practices, and although there is some spill over and interaction, there is also evidence that publics act differently in different settings, especially when contrasting home and work (Chapter 8). Again, hierarchies of worth can be deployed implicitly to value actions in the public sphere such as voting (Chapter 7) as more democratic and potentially more powerful than actions in the private sphere such as consuming (Chapter 4) or having fun (Chapter 5).

Another aspect of spatialisation is how spaces can be linked, e.g. through people travelling for work, to demonstrate how (and how far) environmental practices are able to move through space, to be shared, adapted or resisted by other publics. Such links may be topological rather than cartographic, that is, they may reflect personal networks and professional collaborations rather than proximity on maps: city professionals living in London may have more work contacts in New York than in Manchester, academics in Cambridge may work closely with others in California or Moscow, so that working practices may seem to jump across space.

So as well as the materials, meanings and skills cited by practice theorists (e.g. Shove and Pantzar 2005; Truninger 2011), places also matter: practices are carried by people but also carried through contexts of performance, moving across and between spaces. Practices grow in complexity as they travel, are reinvented and adapted from place to place, from person to person, rather than merely diffusing across space. Addressing the geographies of practice is therefore essential if we wish to analyse more fully how environmental publics and practices co-evolve.

Material and virtual publics

Another aspect of spatialisation is the contrast between online and offline spaces of environmental practice. In this book, I have not devoted a chapter specifically to online practices, but have instead mixed together discussion of offline and online ways of shopping, campaigning, learning and so on in each chapter. Partly this was because online practices often merely adapt and speed up offline practices, such as e-petitions or mass emailing instead of mass letter-writing, for example, changing the technology and perhaps the skills needed but not the meaning of such practices.

But it was also because I wanted to emphasise how environmental publics incorporate the material and the virtual, the physical and the imagined. In the same way that publics use tangible devices (cars, computers, walking boots) to enact environmental practices, they also use intangible repertoires of meaning and expectation and digital technologies of communication and representation. In this sense, 'virtual' means more than merely 'online' or 'digital' – it also refers to the imagined histories, normative expectations and idealised figures (e.g. the 'general public' or 'the ethical consumer') that designers of public participation exercises or retail websites have in mind. This is partly why there have been so many studies attempting to distinguish what people do environmentally by what kind of person they are, that is, to give sociodemographics and other personal attributes correlative or even causal roles in explaining who is more or less environmentally impactful (as mentioned in Chapter 1). And where such studies rely on self-reporting of attitudes and behaviours, there is even more scope for individuals to (re)imagine themselves through drawing on virtual figures and expectations.

So there are some parallels between virtual imagining and online practices, especially where online spaces allow people to conjure their imaginaries as digital avatars. Analysing online gaming in *Second Life* and *World of Warcraft*, Denegri-Knott and Molesworth (2010, p. 113) argued that virtual consumption encompasses both online simulation and offline imagination, and may today be more important than material consumption. In the case of environmental practices, the previous chapters have shown how meanings and ideals matter as much as devices, locations, bodies, energies and other physically measurable elements in shaping environmental publics and what they do.

A good example to emphasise this point is food. We consume food materially by putting it into our bodies, but we also consume food virtually through ideas and associations. From Proust's madeleines to the names of dishes, the sense of taste is only one way in which we engage with food – also significant are smells, sight, touch (texture) and how we think about food, what is taboo, what we expect to find in a dish and what we do not. Another simple example is how we may value a birthday cake that a friend has made for us far more than one bought for us.

So material and virtual practices are not carried out in separate domains, but are intimately bundled together to create and shape environmental publics. Similarly, online and offline practices are often linked: people may browse for products on an internet website but then go into a bricks-and-mortar shop to purchase them, and online browsing and offline shopping both support the virtual imagining of products that goes before the tangible, material use of them.

To take food as an example again, imagined atrocities or environmental damage are often a reason for vegetarians and environmentalists to avoid foods, but these are 'imagined' elements, imagined not because they do not happen somewhere at some time, but because they are not physically evident or detectable in the product itself through human senses, unlike other attributes such as taste or price. Consumers may visualise foods negatively in their minds, drawing on other images, information and stories told about battery chickens, incarcerated calves or salad leaves washed in bleach. Finding out about a type of food (or deliberately *not* finding out about how it is produced for fear of learning something horrible or unpalatable) is part of the experience of consuming that food.

People thus make food virtually and imaginatively through giving it meanings, and make food materially and tangibly in gardens and allotments, kitchens and picnics. And if people encounter a vegetable, say, that they do not know how to cook and/or have never eaten, then this lack of virtuality, their inability to imagine how they might consume it, may prevent them consuming it materially through purchasing and eating it. We can thus regard both resources such as recipes, cookbooks, TV programmes and YouTube videos, whether they are encountered online or offline, all as ways that help consumers to imagine foods into being in their own lives.

Interventions into the virtual often try to conjure environmental meanings into products through story-telling, symbolisation and information about intangible attributes, to encourage certain forms of consumption that are felt to be more environmentally worthy. There are many examples of this being attempted through the certification and ecolabelling of products. As we saw in Chapter 4, a tiny number red-inked onto an eggshell can be used to link that egg virtually to a place, a named farm near a named town run by named people, despite the egg looking exactly like (literally) millions of other eggs laid every year in its shape, colour, nutritional content and flavour. In this case, the virtual attributes that people cannot detect for themselves by encountering the product, such as how it was made, are constructed and communicated using digital technology over the internet. By connecting things with individual farmers and small producers or processors, products may be given 'personality' in people's imagination, as well as ethical and environmental worth, although this is not always successful.

With the growth of internet access, the miniaturisation of mobile computing and the embedding of digital tags into material products, it is increasingly clear how online and offline spaces intertwine in people's daily lives, emphasising how spatialisation diversifies but also links together environmental publics through practices.

Changing practices

A more vexed question is, how do practices change? This is both a conceptual problem that plagues researchers analysing social practice and a practical problem for policy makers and practitioners aiming to change human behaviour in order to reduce the environmental impacts of everyday actions. If we understand practices in everyday life as both made by and making environmental publics, especially through collective routinisation, then how (and where) do we understand the process of change originating? If the lone hero or champion is a deficient model for changing social norms, and I have suggested it is, where can we locate agency? In previous chapters, we have seen how routines can develop, be shared and become entrenched, making the status quo difficult to challenge, but it is harder to understand how new routines are initiated or why some routines are not initiated or fail to spread or become entrenched.

In a way, this requires us to see environmental publics not as the source of agency or the instigators of new practices, or at least not alone: other elements are necessary to enact practices and also to change them, through building assemblages with distributed agency. In this vein, Lee and Motzkau (2013) criticised approaches that distinguish humans, their will to act and their effects on the environment from other factors, calling approaches "anthropocentric" for regarding environmental phenomena (such as climate change) as caused by and therefore fixable by deliberate human action, and "anthropomorphic" for ascribing moral responsibility for causing such events solely to humans. Instead, they argued that environmental phenomena are more correctly understood as "emergent biosocial phenomena", acknowledging that it is difficult, maybe impossible, to identify causal agents or to ascribe responsibility for change to those agents. To put this another way, they argued that to see climate change as a problem caused by human actions that needs to be solved by human actions is inadequate to the point of arrogance.

This argument has the practical consequence of problematizing any attempts to change what publics do in order to lessen their environmental impact, attempts that are critical to much environmental policy today, especially where it relies on persuasion rather than regulation. Shove (2010) particularly criticises the emphasis in environmental policy on persuasion, where policy makers attempt to lever individual choices towards more environmentally friendly options, rather than appreciating the context and infrastructural conditions in everyday life that shape (and often restrict) those choices. As well as suggesting that policy makers pay heed to other theorisations of environmental practices, she argued that policies for public health and planning offer more systemic ways to redirect practices through changing how services such as electricity and transport are provided and how spaces such as cities and domestic houses are designed so as to lessen environmental impacts.

In a similar way, in this book I have emphasised how environmental publics are made through the contexts and consequences of practices, as well as enacting, adapting, moving, resisting and sometimes dropping those practices. Rather than being drivers of behaviour, in this book attitudes are both routes through these processes and consequences of them, as people give meanings to practices, learn from others and shape imaginaries of how they might act through participation and protest. Change is thus integral to an understanding of environmental publics as enacted and performed, but deliberately engineering change cannot be achieved simply by finding the right spot to apply leverage: rather, it is only possible through seeing the wider interrelationships between contexts, meanings, habits, things, places and the environmental publics that are caught up in them.

Summing up

Throughout this book, I have emphasised how practices make publics and publics enact practices, that is, how they co-evolve, and particularly the geographies of this co-evolution in a range of different contexts, from working to voting and playing. In all cases, I examined how the conceptualisation of practices as more or less worthy, more or less powerful, more or less transformative is problematic, because it tends to construct figures of 'the consumer', 'the citizen', 'the manager' onto which practices are projected. These imaginaries are then used to generalise about and often to denigrate publics as more or less politicised, more or less empowered and more or less sustainable, tending to stabilise and restrict how we think about environmental publics.

I have also emphasised the active and continually changing character of environmental publics, as they enact, adapt and are shaped by practices. How publics engage with environmental issues is therefore not predictable as the product of their sociodemographic measurements, nor through other singular characteristics such as the magazines they read or the outdoor recreations that they enjoy. Their views do not change as information is received, nor as they move house or workplace. And environmental publics are also active agents of their own identities and imaginaries, being part of the processes of giving meaning to their own actions and the actions of others. So, focusing on both publics *and* their practices is necessary, but we need also to consider the geographies, both topological and cartographic, implicit in how environmental publics and their practices co-evolve across time and space. Placing the assemblage of things, meanings and skills emphasises the spatialisation of environmental practices and how practices, publics and places together produce new environmental realities.

References

Denegri-Knott, Janice and Mike Molesworth (2010). Concepts and practices of digital virtual consumption. *Consumption Markets & Culture* 13, 2, 109–132.

Devine-Wright, Patrick (2013). Think global, act local? The relevance of place attachments and place identities in a climate changed world. *Global Environmental Change* 23, 61–69.

Lee, Nick and Johanna Motzkau (2013). Varieties of biosocial imagination: reframing responses to climate change and antibiotic resistance. *Science, Technology, & Human Values* 38, 4, 447–469.

Shove, Elizabeth (2010). Beyond the ABC: climate change policy and theories of social change. *Environment & Planning A* 42, 1273–1285.

Shove, Elizabeth and Mika Pantzar (2005). Consumers, producers and practices: understanding the invention and reinvention of Nordic walking. *Journal of Consumer Culture* 5, 1, 43–64.

Truninger, Monica (2011). Cooking with Bimby in a moment of recruitment: exploring conventions and practice perspectives. *Journal of Consumer Culture* 11, 37–59.

INDEX